The Birth of Homo, the Marine Chimpanzee

The Birth of Homo, the Marine Chimpanzee

When the tool becomes the master

Michel Odent

The Birth of Homo, the Marine Chimpanzee:
When the tool becomes the master

First published in Great Britain by Pinter & Martin Ltd 2017

ISBN 978-1-78066-445-3

Index: Helen Bilton

British Library Cataloguing-in-Publication Data
A catalogue record for this book is available from the British Library.

Set in Minion

Printed and bound in the EU by Hussar

This book has been printed on paper that is sourced and harvested from sustainable forests and is FSC accredited.

Pinter & Martin Ltd
6 Effra Parade
London SW2 1PS

www.pinterandmartin.com

Contents

1

Why humanity is at a 'turning point'

We are undoubtedly at a 'turning point' in the history of the relationship between mankind and 'Mother Earth'. At a 2016 international geology conference in South Africa, participants agreed that we are entering a new age: human activities are now the main factor influencing the evolution of earth chemistry. The advent of the 'Anthropocene' epoch is a 'turning point' in the history of our planet.

Some decades ago, comments on the weather were just the usual way to start conversations between farmers, gardeners or lay people. Today the most powerful heads of state meet at conferences to discuss the effect of human activities on the evolution of our climate. A turning point!

According to a WHO air quality model, 92 per cent of the world's population lives in places where pollutants exceed WHO limits. We have reached a turning point beyond which accelerated action is needed: to improve our quality of life we need sustainable transport in cities, solid waste management, access to clean household fuels and cook-stoves, industrial emissions reductions and a shift towards renewable energies. It is an indisputable fact that we have reached the turning point beyond which renewable energy sources must replace fossil fuels.

Since oceans cover most of planet Earth, we must realise the paramount importance of marine pollution. Millions of tons of debris end up in the oceans every year and 60 to 80 per cent of that debris, or 4 to 5 million tons, is discarded plastic litter. Plastic pollution was first noticed by scientists carrying out plankton studies in the late 1960s and early 1970s. We understand today that, in addition to being non-nutritive and indigestible, plastics concentrate pollutants up to a million times their level in the surrounding sea water and then deliver them to the species that ingest them. Here too we have reached the turning point and urgent action is now needed.

Countless other examples of appropriate use of the term 'turning point' can be found if we look at what we know about life on Earth. Another example is the 2012 report by the International Union for Conservation of Nature (IUCN), which listed 3,079 animal and 2,655 plant species worldwide as endangered.

* * *

All these separate 'turning points' are not just concomitant events. They are part and parcel of a major crisis in the history of humanity and planet Earth. To understand the nature of this crisis, we must refer to the previous one, which started about 10,000 years ago in places as diverse as the Middle East, South Asia, Central Asia and Central America. This was when our ancestors started to dominate nature, instead of existing as part of it. Commonly called the 'Neolithic revolution', the crisis had several facets: domestication of plants and animals through agriculture and animal husbandry; socialisation, regulation and control by the cultural milieu of human physiological processes related to reproduction (genital sexuality and childbirth); and canalisation of the universal transcendent emotional states. It is essential to realise that Homo too was 'domesticated' to a certain extent.

Today, after a recent and spectacular acceleration of our domination of nature, it is as if we have reached the edge of a precipice. To survive, humanity must urgently change direction – if we do not, we will plunge into the abyss. To characterise such a counter-revolution, which might prolong the survival of mankind, we need a term that is the antithesis of 'domination' (of nature). I suggest 'symbiosis', since it literally means 'with life'. A 'symbiotic revolution' would have two main facets. First, increased 'networking' with other living creatures, particularly microorganisms, which are the foundations of all ecosystems. Second, improving our understanding of the laws of nature in order to work with them, instead of neutralising them.

Nobody would dare to predict the effects of such a hypothetical counter-revolution. We can only interpret our reasons to focus on childbirth, this critical phase in human development highly influenced by cultural conditioning.

If there is no paradigm shift without a renewed vocabulary, it is worth noticing that we are already at a turning point in the evolution of language: the term 'turning point' is itself a neologism.

2

A turning point in our understanding of the human condition*

Since the current crisis in the history of planet Earth is related to human activities, our first step is to focus on the countless particularities of our own genus Homo in the framework of mammal species. For each particularity, even the most mysterious, multiple interpretations have been suggested in terms of evolutionary advantages and adaptation to different environments. I propose that applying one simple rule – largely inspired by advances in physiological sciences – can introduce a unifying theoretical framework. When a trait is mysterious, and is apparently specific to humans, we must look at what we have in common with mammals adapted to the sea.

A gigantic and highly developed brain

The main trait that makes human beings different from other mammals is their enormous brain mass in relation to the size of the body, which is usually quantified in

---

* *Author's note:* Information given in small type between brackets is not essential for an understanding of the principles of this chapter, but may be of interest to those who are familiar with biochemistry.

terms of 'encephalisation quotient'. Compared with the other members of the chimpanzee family, from which we probably separated about six million years ago, and although the genetic difference between humans and chimps is less than two per cent, our encephalisation quotient is mysteriously three times higher. If we follow the general rule I proposed above, we discover that the encephalisation quotient of bottlenose dolphins is roughly twice that of common chimpanzees and bonobos. Mammals adapted to the sea generally have higher encephalisation quotients than their cousins on land.

Another mysterious aspect of human nature is the association of a gigantic and complex brain with an enzymatic system that is not very effective at making a molecule of fatty acid commonly called DHA. This molecule, essential to feed the brain, is abundant and pre-formed only in the seafood chain, which suggests that humans are programmed to have this molecule in their diet. In practice, this means access to seafood. DHA accounts for 50% of the molecules of fatty acids that are incorporated into the developing brain. (Land-based food can easily provide the parent molecule of the omega-3 family [18 carbons and 3 double bonds]. To make DHA [22 carbons and 6 double bonds] the enzymatic system must elongate and desaturate the parent molecule). In general, the study of enzymatic systems is an effective way to analyse the main characteristics of animal species. Viewed in this way Homo can be seen as a member of the chimpanzee family adapted to the coast.

We can also learn about human nature by focusing on the most common nutritional deficiency. All over the world, human beings struggle to meet their need for iodine, unless they have regular access to seafood. It is an important issue, since having an enormous highly developed brain implies daily iodine requirements. Iodine is a 'brain selective nutrient' because of its essential role in thyroid hormone production, which, in turn, is needed for normal brain development. Iodine is scarce on the earth's surface because over hundreds of millions of years it has been washed away by rain and glaciation and transported from the terrestrial crust to the sea. Seawater contains 10 to 200 times more iodine than fresh water.[1] In the sea, algal phytoplankton, the basis of the marine food chain, act as a biological accumulator of iodine.

Iodine is the only nutrient for which governments legislate supplementation, so iodination of table salt is mandatory (the iodine in sea salt disappears during the process of desiccation in salt marshes). In spite of widespread legislation and many public health strategies (such as dripping iodine in the water of Chinese irrigation ditches), iodine deficiency remains a common nutritional deficiency on a global scale. It is the leading cause of preventable intellectual disability. It is difficult for humans to obtain sufficient iodine if their diet does not contain seafood. Further, some common agricultural practices exacerbate the problem: the milk of cows fed with rapeseed meals rather than grass, for example, has an iodine antagonist effect.[2]

These issues are particularly serious for women of childbearing age, who need increased amounts of iodine during pregnancy and lactation. According to the ATA (American Thyroid Association), women should take a daily supplement (150µg) of iodine before conception, and during pregnancy and lactation. According to a British study, a daily supplement of iodine when the mother was pregnant was associated with an average of an 1.22 point increase in IQ.[3] In general, public health organisations worldwide must establish strategies to satisfy such basic universal human nutritional needs, which are suggestive of adaptation to the coast.[4,5]

This quasi-universal need to supplement iodine in populations that do not consume seafood on a regular basis can usefully be looked at alongside our tendency to recommend small daily doses of aspirin to keep healthy, which is becoming ever more widespread. It is a long story, since 'aspirin', in the form of leaves from the willow tree, has been used for thousands of years. There is even evidence that tens of thousands of years ago Neanderthals living in El Sidron cave, in Northern Spain, were consuming the leaves of poplar trees, also a source of 'aspirin'.[6] Now nearly all people over 50 are advised to routinely add aspirin to their diet. Questions have also been raised about the possible beneficial effects of the routine prescription of aspirin in all pregnancies(!).[7] One of the main modes of action of aspirin was revealed by John Robert Vane in 1971, following the discovery of 'prostaglandins' as local hormones produced in the body.[8] Aspirin suppresses the production of 'eicosanoids' (prostaglandins and related compounds) by a mechanism of enzyme inactivation. It tends to correct a widespread current dietary imbalance, characterised by a comparative excess of fatty acids of the omega 6 family (as opposed to omega 3). (The effect of this dietary imbalance is an excess of synthesis of the family of eicosanoids derived from arachidonic acid, a 20-carbon omega 6 fatty acid with 4 double bonds. These eicosanoids are pro-inflammatory and, by facilitating platelet aggregation, are also clotting agents.)

The important point is that production of eicosanoids derived from the omega 6 family is minimised by regular consumption of seafood, which is rich in the long-chain fatty acids of the omega 3 family (EPA and DHA). (There is a mechanism of enzymatic competition. EPA gives rise to eicosanoids that often have lower biological potency than those produced from arachidonic acid, and EPA and DHA give rise to anti-inflammatory and inflammation-resolving agents).

The beneficial effects of aspirin can thus be considered suggestive of the importance of seafood in human nutrition. It is highly significant that daily consumption of aspirin has no detectable preventive effects in the particular case of Japan, a country with high levels of seafood consumption.[9]

It is probable that, in the near future, our knowledge of sea mammals will shed light on the particularities of other human metabolic pathways. (For example, this might be the case for the 'sialic acid family', which includes derivatives of the sugar neuraminic acid. It has been demonstrated that the human brain has significantly more sialic acid than the brains of other mammals. It is not mysterious if we refer to the comparative studies – done at SeaWorld San Diego – of the urine of humans, orcas and dolphins.[10])

Two intriguing particularities of human births[11]

It is commonly said that only the skin of human newborn babies is covered by vernix caseosa (literally cheesy varnish), a greasy white substance secreted by the sebaceous glands during life in the womb. In many cultures the vernix was denied any role and routinely wiped away. But its importance becomes clearer if we look to mammals adapted to the sea to explain this mysterious and overlooked human characteristic.

Don Bowen, a marine biologist from Nova Scotia, Canada, observed that the pups of seals also have vernix. Interestingly, he noticed that harbour seals, which swim with their mothers within minutes of being born, have more vernix than other seals, which do not swim for at least 10 days. It appears that although approximately 80 per cent of vernix is water, it still has high viscosity, suggesting that its water is contained within a highly structured state conferred by an abundance of water-filled foetal 'corneocytes'. These foetal corneocytes act as 'cellular sponges' that prevent water from moving across the skin, whereas sebaceous lipids, including squalene, provide a hydrophobic barrier. Vernix is so rich in squalene that a measure of its concentration in amniotic fluid had been suggested as a test to detect the effects of postmaturity.[12] It is noticeable that squalene is more abundant in the aquatic animal kingdom than among other animals.

Thus vernix caseosa could be interpreted as a transitory protection against immersion in salty, hypertonic water. We should remember that vernix caseosa is a common point between Homo and seals, but is unknown among land mammals.

It is also intriguing that, apparently, eating the placenta has never been instinctive in our species. If 'placentophagy' had been a common behaviour at any time in the history of humanity, we should find hints at this behaviour in myths, legends, and reports from pre-literate and pre-agricultural societies. I know of women who have reached a very instinctive state of consciousness in the perinatal period, behaving as if 'on another planet' and overcoming a great part of their cultural conditioning. Yet none of them has ever expressed a tendency to bring the placenta towards their mouth. Modern women who consume pieces of the placenta have been inspired by theories, such as the theory that it might prevent postnatal depression. Scientific interest in the placenta has inspired such theories, leading to a form of human placentophagy based on rational considerations. For example, the discovery by Kristal of a placental substance that makes endorphins more effective ('Placental Opioid-Enhancing Factor') could be seen as a justification for placentophagy in our species.

However, we should avoid the conclusion that eating placenta is an innate human behaviour.

Exploring placentophagy is important since all land mammals eat the placenta. If, as we suspect, eating the placenta has never been instinctive among our ancestors, this would be another common point with sea mammals, including cetaceans and seals. Interestingly, camels are the exception among land mammals: they never eat the placenta. Camels have another particularity among land mammals: like Homo and sea mammals they have kidneys with multiple 'pyramids'.[13] Since camels consume salty plants and drink water from salty ponds after giving birth, and since sea mammals also have easy access to hypertonic salty substances, we can see that placentophagy and special kidney structures might be correlated with an urgent need for specific nutrients, particularly minerals, in the period following birth. It is as if placentophagy and the usual kidney structures are features shared by mammals that do not have access to hypertonic salty substances after giving birth. Can camels from the desert help us to understand our common points with sea mammals?

Countless other intriguing human features

Many other intriguing human features have been considered by the pioneers of the so-called 'aquatic ape hypothesis'. Several of them were included in the historical lecture Sir Alister Hardy delivered in March 1960 at a conference organised by the British Sub-Aqua Club in Brighton.[14] Elaine Morgan looked at these features in detail and explored them in her books,[15,16,17] and also at events she organised (she invited me to participate in a conference of the British Association for the Advancement of Sciences in Southampton in 1992, and in a study day in San Francisco in 1994).

Books by nutritionists Michael Crawford[18] and Stephen Cunnane,[19] and also the recent collective book *Was Man More Aquatic in the Past?* are other important steps in the history of the theory.

The point is not to analyse in detail each intriguing human characteristic, but to give a list of them in order to reinforce the general interpretative rule we are using. None of these features alone has the power to support the theory, but bringing all of them together can. Meanwhile, keep in mind another simple rule: when two species are genetically close to each other, morphological and behavioural differences are usually explained by adaptation to different environments. Conversely, animals that are not genetically related can have many common points if they are adapted to the same environment. For example, New World vultures and Old World vultures look so similar that they have been confused by zoologists. In fact, in the age of genetics, we now know that the American birds are genetically related to storks, while the Old World birds are related to hawks.

- Nakedness is certainly one of the most striking human traits: it is shared with aquatic mammals in general, and also with land mammals such as elephants and rhinos, which are in fact very aquatic and need mud to protect their naked skin.
- A layer of fat under the skin is another feature shared with sea mammals. It may be the equivalent of blubber, the layer of vascularised fat tissue found under the skin of all cetaceans (whales, dolphins, and porpoises), all seals and all sea cows.
- The general shape of our body and our upright stance are compatible with life on the coast: it is easier to keep vertical when moving in shallow water. Human babies, for example, can stay erect

and walk in water before being able to walk on dry land. As Sir Alister Hardy wrote: 'It seems to me likely that Man learned to stand erect first in the water and then, as his balance improved, he found he became better equipped for standing up on the shore when he came out'.

- Body temperature control through the loss of sweat has often been considered to be a biological blunder. It is a costly mechanism, depleting the system of large amounts of water, sodium, and other essential elements. New interpretations become possible if we think of the human being as a primate adapted to environments where water and minerals are available without restriction.
- A prominent nose is a feature shared with the proboscis monkey, which looks very human. It lives in the coastal wetlands of Borneo and is an excellent distance swimmer. It uses swimming to escape the cloud leopard.
- Still in the area of the upper respiratory tract, our larynx is low, which gives us the ability to breathe through our nose or our mouth. Sea lions and dugongs are also characterised by a low larynx.
- The human vagina, like that of sea mammals, is long and oblique, and is protected by a hymen.

Some similarities between human beings and sea mammals are bizarre and apparently uninterpretable. However, it is worth mentioning them. Menopause, and prolonged life after reproduction, is often considered a specifically human feature.[20] However, if we look at sea mammals, we learn that female killer whales and short-finned pilot whales spend two-thirds of their lives not birthing any offspring.[21,22] Female killer whales typically start reproducing at age 15, and stop in their 30s and

40s. Yet they can live to be more than 90. The point is that, although these particularities of sea mammals are documented, they are never taken into account in the countless theories about human menopause.

The differences between human hands and those of the other members of the chimpanzee family are also intriguing. The main difference is a triangle of skin between thumb and forefinger. This triangle of skin, which looks similar to the webbing of a duck's feet, is compatible with adaptation to water. The same kind of observation can be made regarding our feet: the big toe is joined to the others in man, but is separated in chimpanzees.

This leads me to mention one of the most common congenital abnormalities (or particularities) human beings can have, namely webbing between the second and third toes. This is highly significant because a congenital abnormality that takes the form of adding a feature usually means that the feature was there for a reason during the evolutionary process.

A cultural blindness challenged by genetics

For hundreds of years, the countless philosophers, scholars and scientists who pronounced on human nature did so without seeing that man has many characteristics suggestive of an adaptation to the seashore. Why was there such a cultural blindness?

Consider the particular case of experts in human evolution. Until recently, fossil hunters were the only ones with authority to propose theories on human origins. Today we can evaluate the limits of this perspective:

- 20,000 years ago, due to the volume of ice on land, sea level was still approximately 130 metres lower than it was when the Neolithic revolution started, about 10,000 years ago. This implies that if most human beings were living along the coasts during the last glacial period, which occurred from 110,000 to 10,000 years ago, only the fossils of the minority of humans who were living inland will ever be found.

- Only hard parts of the body, such as the teeth, skull, jaw and limb bones can be preserved and found in fossils. In addition, fossils are often distorted by earth movements during their long burial. As many adaptive characteristics concern the soft tissues only, it is impossible, for example, to distinguish the fossils of a tiger from the fossils of a lion.
- When human fossils are found, it means that the geological conditions are absolutely exceptional. A fossil discovery belongs to the realm of improbability. First, to become fossilised, human remains must survive the normal process of decay that returns flesh and bone to dust. Then comes the next highly unlikely event: just as the bone is exposed, but before it can be eroded, a skilled palaeontologist must happen along. The British science writer Tim Radford has concisely clarified what we need to understand: 'the surprise is not that there are not more fossil remains; the surprise is that there are any at all'. The French geologist Jacques Varet, who has a good knowledge of East Africa, enumerated all the geological conditions that have to occur together to give a hominid fossil even a tiny chance of ever being detected. East Africa is so exceptional from a geological point of view that it is not surprising that fossils of hominids living 2 or 3 million years ago were found there.[23] But we should nevertheless be

> cautious about suggesting that the emergence of humanity occurred in this part of the world.

It is in this context that geneticists are gradually becoming crucial authoritative experts in human evolution, able to challenge the dominant assumptions and theories. From this perspective, it appears today that it is mostly by starting from coastal bases that our ancestors have colonised the whole planet. This is more and more obvious when considering the colonisation of the Pacific Rim in general, and the American continents in particular. The genetic perspective, complemented with what we know today about low sea levels during the last ice age and the very ancient techniques of canoeing, offers a new way to evaluate the special relationship between Homo and the sea.

The canoes of Neanderthals and Sapiens

It appears today that one of the cradles of human civilisation has been the prehistorical lowlands of the Southeast Asian peninsula, that our ancestors probably reached more than 50,000 years ago. This area, commonly called the 'Sundaland', was above sea level during the last ice age. It was twice the size of India, and included what we now call Indo-China, Malaysia and Indonesia. From there, some of our ancestors migrated towards the South, reaching the ancient continent of Sahul (which is now divided into Australian mainland, New Guinea and Tasmania). Others migrated towards the North, reaching the Japanese Archipelago about 40,000 years ago.

The 21st century has also brought a new understanding of the migrations towards the American continents. Until recently, according to theories based on archaeological data, the ancestors of the indigenous cultures of the American continents appeared in what is now New

Mexico, where they developed the 'Clovis culture'. They were supposed to have reached the North American continent through an ice-free corridor that extended from Alaska to Montana. We have recently learned from Danish geneticist Eske Willerslev and his team that life came to the ice-free Canadian corridor too late to sustain this theory. Studies of plant and animal genetic material indicate that the passageway became habitable nearly 1,000 years after the formation of the Clovis culture.[24] So there is now an accumulation of reasons to favour the theory of a coastal migration route. This is an unexpected opportunity to consider the marine aspect of human beings. The use of canoes is probable.

There are several other reasons to change our way of thinking about the earliest inhabitants of the Americas. According to radiocarbon dating, human beings were living in Monte Verde, on the Chilean Pacific coast, in at least 14,800 BC... about 13,000 kilometres from Alaska![25] Furthermore, genetic studies have demonstrated that some Amazonian Americans descend partly from a Native American founding population with ancestry more closely related to indigenous Australians, New Guineans and Andaman Islanders than to any present-day Eurasians or Native Americans.[26]

In the current scientific context, a human migration towards the coasts of South America via the southern part of the Pacific Ocean is plausible, and even probable. We must keep in mind that 20,000 years ago, when sea levels were more than 100 metres lower than today, there were countless islands between Polynesia and the Chilean coast. Some of them still exist, for example: Pitcairn Island, Easter Island, Sala y Gómez, the Desventuradas Islands (San Félix and San Ambrosio), Alejandro Selkirk and Robinson Crusoe. We must also keep in mind that canoeing was a widespread prehistoric human activity.

This fact alone is a sign of the deep-rooted relationship between Homo and the sea. Nobody knows when our ancestors started to make canoes in the Pacific area, but it is probable that hominids even older than Neanderthals were already seafarers. One-million-year-old stone tools have been found on the Indonesian island of Flores, suggesting that primitive Homo erectus had probably crossed the sea.[27]

There is also evidence that the Neanderthal variety of Homo could reach Mediterranean islands by canoeing.[28,29,30] Neanderthals lived around the Mediterranean from 300,000 years ago. Their distinctive 'Mousterian' stone tools are found on the Greek mainland and, intriguingly, have also been found on the Greek islands of Lefkada and Kefalonia. According to George Ferentinos, from the University of Patras, Greece, the sea would have been at least 180 metres deep when Neanderthals were in the region.[31] In 2008 Thomas Strasser, from Providence College in Rhode Island, found similar stone tools on Crete, which he says are at least 130,000 years old. Crete has been an island for some 5 million years and is 40 kilometres from its closest neighbour. There is food for thought in Thomas Strasser's comment: 'Early hominids may have used the seas as a highway, rather than seeing them as a barrier'.

This parenthesis about Neanderthals can help us to accept the concept of migration through the South Pacific ocean. We must also take into account how important a part was played by birds in the guidance of early navigators in uncharted seas.[32] It is significant that the Rapa Nui name for Sala y Gómez island means 'Bird's islet on the way to Hiva'. Even in historical times, explorers were still guided by birds. Vicente Pinzón, the Spanish navigator who sailed with Christopher Columbus on their first voyage to the New World, was quoted as saying: 'Those birds know their business'.

Last bombshell: a 130,000-year-old archaeological site has been found in southern California![33]

The story started with the discovery of fossils of mastodons. Before knowing more, we must oscillate between curiosity and scepticism… Let us keep in mind that, in the near future, the detection of human genetic material in sediments will probably be possible, when fossils (bones, teeth, etc.) are not available.[34]

Know thyself

After combining the physiological perspective, the genetic perspective, and what we now know about the routes our ancestors probably followed to colonise the whole planet, it seems clear that the genus Homo has the characteristics of a member of the chimpanzee family adapted to the land-sea interface. Those who still need other perspectives can observe toddlers in paddling pools and meditate on the effects of the advent of paid-for holidays, a phenomenon that suddenly appeared during the 20th century. Millions of human beings now concentrate on beaches and spend time watching the waves. Other bizarre aspects of human behaviour can also give rise to new interpretations. For example, we may be surprised by the current tendency to confuse 'natural childbirth' and 'waterbirth', while seals go on giving birth on dry land.

To properly examine human nature, and circumvent the cultural blindness that prevents us from proper understanding of our humanity, we must analyse the ways of thinking that have developed since the advent of agriculture, animal husbandry and other aspects of the domination of nature. While the five continents were originally colonised by human beings living by the sea, the Neolithic revolution started among human groups that had migrated inland and were obliged to adapt to life away from the ocean. It began in places as far apart as the 'Fertile Crescent', southwest Asia, China, the Carpathian Basin in Europe, the Ethiopian highlands, the Nile river valley, the Andes mountains in South America, the Mexican highlands and also the highlands of Papua New Guinea. All these places are far from the sea.

Today, when our domination of nature has reached extreme limits, we must reconsider our strategies for survival. If we are to find a new way of understanding our basic human needs, we must take the aphorism 'know thyself' as an incitement to explore the genus Homo from novel perspectives.

3

A turning point in our understanding of human birth

Just as we are learning about human nature from new perspectives, we are also at a turning point in our understanding of human births. Until now, the focus has been on mechanical difficulties. Countless textbooks have reproduced drawings showing the size and the shape of the foetal skull in relation to the maternal pelvis as a way to explain why the birth process cannot be easy in our species. If the main reasons for difficulties were mechanical, how do we explain the fact that, occasionally, women who are not special, from a morphological perspective, have their first baby easily within a few minutes, while others need a caesarean section after one or two days of tough labour?

When the tool becomes the master

Once more, our point of departure should be the main attribute of the 'marine chimpanzee', in comparison with other mammals in general and other primates in particular: its gigantic brain. The human brain includes a highly developed 'new brain', or 'neocortex'. The neocortex has usually been presented as a tool at the service of vital physiological functions, providing information about space and time, and facilitating communication. However, recent work has shown that we should consider the capacity of the neocortex to occasionally prevent or inhibit physiological functions. Now it seems that the concept of neocortical inhibition may be key to understanding human nature. There are countless published studies about excitatory neurons, inhibitory neurons and neurons specialised in disinhibition, but to understand the importance of cortical (and neocortical) inhibition we do not need to refer to them in detail. We just need to understand that brain excitation cannot be permanent, and that there is a biological mechanism that can slow down or stop neocortical activity in some specific situations. It is as if there are physiological accelerators and also physiological brakes. The concept of cortical inhibition was proposed in the second half of the 20th century, after Eugene Roberts revealed the presence of GABA (gamma-aminobutyric acid) in the tissue of the vertebrate central nervous system in 1950.[1] It was soon found that GABA is the major inhibitory neurotransmitter.[2]

The concept can be illustrated using examples of human physiological functions that are easily inhibited by neocortical activity. The sense of smell, apparently weak among adult humans, is one example. An ingenious experiment in Israel elegantly demonstrated that the human sense of smell is improved following alcohol consumption: it is well known that alcohol reduces inhibition.[3] It is highly significant that olfaction appears to be an important physiological function in newborn babies until a certain degree of neocortical development has taken place. In the 1970s I observed that, immediately after birth, it is mostly through the sense of smell that the baby is guided towards the nipple.[4,5] Since that time, there has been intense curiosity, in scientific circles, about the function of the sense of smell in early infancy.[6 to 14] One of the reasons for this curiosity is that the sense of smell becomes gradually weaker in the weeks following birth. Neocortical inhibition may be the best explanation of this phenomenon.

Consider also the ability to swim. The turning point in our understanding of human swimming behaviour came in 1939 when Myrtle McGraw published the conclusions of her studies.[15] She observed and filmed 42 babies during the first two and a half years of their life. She repeated observations of the same babies at different intervals, making a total of 445 observations. This enormous accumulation of data laid the foundations for our current knowledge of aquatic behaviour at different phases of human life. From her observations we can conclude that the capacity to adapt to immersion and make coordinated swimming movements when submerged disappears at around the age of three months, when the neocortex reaches a certain degree of maturity.[16]

It is also worth recalling that in all human societies, even those where genital sexuality is comparatively free, couples isolate themselves to make love. Such a universally accepted need for privacy in specific situations indicates a deep-rooted understanding of an essential aspect of human nature.

The concept of neocortical inhibition is not commonly mentioned by medical practitioners, apart from psychiatrists in the particular case of pathological conditions associated with culturally unacceptable behaviours, where there is evidence for impaired inhibition. However, the concept is understood in an empiric way by some clinicians. For example, a urologist specialised in prostatic diseases will never ask a man to urinate in front of him to evaluate the power of his urine jet.

It is artificial to separate the issue of neocortical inhibition from the issue of 'primitive reflexes'. These reflex actions are exhibited by normal infants, but disappear after a few weeks or months and are not exhibited by normal adults. The 'rooting reflex', which enables a newborn baby to find the breast during the hour after birth, is one example.[4,5] From a theoretical perspective, it is important to note that older children and adults with pathological conditions such as cerebral palsy may retain these reflexes. Furthermore, the reflexes may reappear in neurological conditions such as dementia, strokes and after traumatic lesions. It is also significant that primitive reflexes can reappear in normal elderly people: this is an obvious sign of the physiological ageing of the neocortex and its declining inhibitory power.[17]

Birth physiology as a chapter of brain physiology

When I was a medical student in the middle of the 20th century, childbirth textbooks focussed on the pelvis, the uterus and the perineum. If we want to create the right conditions for women to continue to give birth in the future, we must change this focus and accept that birth physiology is first and foremost a chapter of brain physiology. This is why one of the prerequisites for the advent of a new paradigm, after thousands of years of socialisation of childbirth, is to constantly refer to the main particularity of Homo.

Even during the second decade of the 21st century, it is still possible for us to examine nature's solution for making human birth possible and sometimes easy. Some people still know that when a woman gives birth easily by herself, without any pharmacological assistance, she may cut herself off from our world, forget what she has been taught, forget her plans and behave in a way that is usually considered unacceptable, for example screaming or swearing. Some women find themselves in the most unexpected, often mammalian, primitive quadrupedal postures. Interestingly some women in full labour complain of odours nobody else can perceive: this is an example of the kind of reduced neocortical control that appears to be necessary for easy birth in humans. Viewed in this way, it is easy to observe and summarise the basic needs of a labouring woman: she needs to feel protected against all possible stimulations of her neocortex. Since language is a powerful neocortical stimulant, it is easy to reach the conclusion that silence is a basic need.

The effects of light on the birth process were not taken seriously until recently, when it was found that melatonin, the 'darkness hormone', is an essential birth hormone. Studies of the interactions between melatonin and other brain mediators offer a promising avenue for further research. The effects of melatonin as an inhibitor of neocortical activity are already well understood.[18,19] When considering the effects of melatonin, and therefore light, on human parturition, we have to deviate from the concept of neocortical inhibition and look to recent advances in research into peripheral effects. It is now established that there are melatonin receptors in the human uterus, and that melatonin works together with oxytocin to enhance contractility of uterine muscle.[20 to 24] It seems that melatonin is an important hormonal agent in human birth. This is confirmed by the significant amount of melatonin in the blood of neonates, except those born by pre-labour caesarean.[25] The importance of these findings is clear when the protective anti-oxidative properties of melatonin are taken into account. In the age of artificial lights, improving our understanding of melatonin release and melatonin properties is important. It is already well established that short-wavelength light (in practice 'blue' light) suppresses melatonin. This fact has been underlined by Cornelia Winner, who found that the lamps in conventional delivery rooms usually emit light rich in the blue part of the spectrum.[26]

It is likely that as our understanding of birth physiology develops, there will be spectacular practical implications. We will attach greater importance to the empiric knowledge of midwives like Ingeborg Stadelmann in Germany, who noticed the beneficial effects of light in the red-orange part of the spectrum.[27] Might we one day consider it rational to give birth by the light of a candle?

In a nutshell, we must remember that all attention-enhancing situations stimulate neocortical activity and therefore inhibit the birth process. 'Feeling observed' is a typical example of this kind of situation: privacy appears as a basic need during labour. The perception of a possible danger is another example: feeling secure is also critical.

I suggest that pregnancy and all phases of labour should be looked at in the light of the concept of neocortical inhibition. As a practitioner familiar with prenatal care, I learned that information transmitted to women in late pregnancy may not be easily retained. This 'memory loss', as a facet of modified intellectual activity, has been scientifically evaluated.[28,29] It can be seen as a physiological preparation for childbirth, with discreet preliminary signs of reduced neocortical power. During the first stage of labour, if there is no incidental interference, women will often become increasingly indifferent to what is happening around them, as if they are gradually cutting themselves off from our world.

A critical phase of labour is preparation for what we have called 'the foetus ejection reflex'. The labouring woman may suddenly talk nonsense, and seem as if she is 'on another planet'. Often what she says may be a short and explosive expression of fear, with a reference to death.[30] After thousands of years of socialisation of childbirth and cultural conditioning, it is exceptionally rare for an authentic foetus ejection reflex to take place.

The term 'foetus ejection reflex' was coined by Niles Newton in the 1960s when she was studying the environmental factors that can disturb the birth process in mice.[31] Twenty years later, with her support,[32] I suggested that we save this term from oblivion. I was convinced it could be key to facilitating a radically new understanding of the process of human parturition.[33,34] When the neocortex is at rest, we have more physiological similarities with mice (and other non-human mammals). A foetus ejection reflex implies a very short series of irresistible and powerful contractions, without any room for voluntary movement. At the very moment of the birth, mothers are typically in an ecstatic state; they may need time to realise that the baby has been born and to take it in their arms. Anecdotal reports confirm how critical these very first minutes are. Once, a mother told me that when she first made eye contact with her baby, it was as if she had seen the whole universe.

This critical phase, including the foetus ejection reflex and the initial interaction between mother and baby, is easily interpreted in scientific terms. An accumulation of data describes the spectacular hormonal upheaval during the minutes before and after birth. The team headed by Kerstin Uvnäs Moberg has demonstrated that, just after giving birth, a mother has the capacity to reach a peak of oxytocin that is even higher than for the delivery itself.[35] This peak of oxytocin is vital as it is necessary for the safe delivery of the placenta with minimal blood loss, and also because oxytocin is the main love hormone. It is undoubtedly associated with high levels of natural morphine and prolactin. Noting that the level of 'adrenaline' can return to normal as early as three minutes after birth, we can assume that a human mother may have an experience which is similar to that of an orgasm. We might present the minutes following birth as an interaction between two human beings who, for different reasons, behave with reduced neocortical control.

Challenging thousands of years of tradition

By examining childbirth in the light of modern physiology, we are challenging thousands of years of cultural conditioning. We present the birth process as an involuntary process under the control of primitive brain structures we share with other mammals. As a general rule, one cannot help an involuntary process, but one can identify inhibitory factors. From a practical point of view, in terms of basic needs, the key word is 'protection'. Since our point of departure has been the concept of neocortical inhibition, protection against language is given first place. If our point of departure had been the concept of adrenaline-oxytocin antagonism, we would have first mentioned protection from scary situations and low ambient temperature to explain that when mammals in general release emergency hormones of the adrenaline family, they cannot release oxytocin, the main birth hormone.

To present 'protection' as a keyword is a preliminary, simple and concise way to challenge tradition. Since the beginning of the socialisation of childbirth, as an aspect of the domination of nature associated with the 'Neolithic revolution', the basis of our cultural conditioning is that a woman does not have the power to give birth by herself. She needs some kind of cultural interference. This dominant paradigm went through many phases, from the advent of midwifery and the most deep-rooted perinatal beliefs and rituals, up to the current masculinisation and medicalisation of the birth environment. Current keywords are eminently disempowering. They always focus on the active role of somebody else other than the mother and baby, the two obligatory actors in the birth drama. They are variants of the concepts of helping, guiding and controlling. The terms 'coaching', used by groups promoting 'natural childbirth', and 'labour management', used in medical circles, imply the intervention of an expert, while the term 'support' suggests that to give birth a woman needs energy brought by somebody else.

When the term 'protection' takes over from our current keywords, we'll have reached a turning point in our understanding of human birth.

4

A turning point in the history of birth preparation

The concept of 'birth preparation' arose in the middle of the 20th century. 'Methods' of preparation proliferated. Paradoxically, they were popularised by groups promoting 'natural childbirth', as if the words 'methods' and 'preparation' were easily compatible with the word 'natural'. The enormous popularity of these methods is not surprising, since they chime with our cultural conditioning: claiming that a woman needs to 'learn' to give birth and needs a guiding expert during labour is just another aspect of our domination of nature after thousands of years of socialisation of childbirth.

Let us recall the words of authoritative pioneers in birth preparation. According to Fernand Lamaze, the father of the so-called Lamaze method, a woman has to learn to give birth in the same way that we learn to speak, or to read. The words of the American obstetrician Robert Austin Bradley were similar. He compared a pregnant woman to a woman who is given nine months' notice that she will be thrown into deep water. Such a woman would use those nine months to learn how to swim. In the same way, a pregnant woman should learn how to give birth.

Physiological preparation

Not only has this chapter in the history of birth reinforced our cultural conditioning, but it has also diverted interest from important aspects of birth physiology in our species. While the focus was on preparation by experts, there was no interest in what we might call 'physiological preparation'. The point is that the signs of this 'physiological preparation' are well known, but could not be taken seriously until the concept of neocortical inhibition had been assimilated.

It has long been noticed that, at the end of their pregnancies, many women are not as mentally sharp as usual. They report memory loss. Their interests change. Their need for socialising may be reduced and reoriented. They tend to restrict or avoid social interactions that are not related to their 'primary maternal preoccupation'. This term, coined by British paediatrician and psychoanalyst Donald Winnicott, is highly significant. Winnicott theorised that, in the last trimester of a pregnancy, through a process that has evolved for its 'survival value', the mother starts to become more intensively aware of, and concerned about, her baby.[1] The 'primary maternal preoccupation' has been illustrated in the past by artists. There are various paintings of Mary, who is pregnant with Jesus, visiting her cousin Elisabeth, who herself is expecting John the Baptist. In several of these paintings, particularly a German one dated 1300, the babies in the womb are visible. This is a clear sign, transmitted by the artist, that these two mothers are mostly preoccupied by their unborn babies and forgetting the rest of the world.

At a time in our history when many women have a

professional occupation involving intellectual activity, we often focus on maternal memory loss. Empiric knowledge is confirmed by scientific evaluations of deficits of several kinds of memory, particularly 'prospective memory', which is the capacity to remember to perform an action at the right time.[2] It is usual to provide advice to 'combat' this condition and to reduce its 'severity': take vitamins, do more exercise, drink plenty of fluids, rebalance your diet. Another aspect of this 'physiological preparation' is a reinforced sense of smell occasionally noticed by pregnant women.

Should pregnant women live in peace?

It is now easy to understand the 'primary maternal preoccupation' that starts before the birth of the baby. It can be seen as the preliminary phase of the reduction of neocortical activity that makes birth possible in our species. It is a component of 'physiological birth preparation'.

There are many other unexplored aspects of the transitory personality changes in the period surrounding birth. For example, it would be worth studying in depth what is called 'motherese', the simplified and repetitive type of language, with exaggerated intonation and rhythm, used by mothers when speaking to their newborn babies. And what about the roots of the universal lullabies?

By using MRI (magnetic resonance imaging) techniques, a joint Spanish and Dutch team has clearly demonstrated that the end of pregnancy is associated with reductions in grey matter volumes, particularly in brain regions involved in social processes.[3] The publication of this study is a turning point with huge practical implications: does it mean that pregnant women should live in peace?

5

A turning point in the evolution of brain size

Until recently, biologists thought evolution was too slow a process to be observed in a human lifetime. Today, from studies of the adaptation of mammals and birds to an urban environment, we are learning that spectacular transformations of species may be fast, and that non-genetic factors have been previously underestimated. As a typical example of rapid transformations related to modified lifestyles, consider the studies looking at plasma melatonin concentrations among blackbirds, which suggest the powerful effects of artificial light on urbanised animals.[1]

In this scientific context there are reasons to examine the particularities of Homo. Some fast transformations are already conspicuous and well documented, such as height. When we consider that Homo is characterised by a huge brain mass, the routine evaluation of head circumference at birth (in relation to birth weight and time spent in the womb) may provide some clues about possible evolution in this area. It has been confirmed that head circumference is a valuable index of brain weight in the newborn.[2] Let's look at three aspects of modern lifestyles that may significantly influence the evolution of head circumference at birth: caesarean section, nutritional factors and environmental pollutants.

Caesarean section

It is commonly accepted that since our ancestors diverged from the other members of the chimpanzee family, there has been a tendency towards a gradual increase in brain volume. An upright posture appears to be a prerequisite for brain development. We can carry heavy weights on our heads when we are upright: mammals walking on all fours cannot do the same. Until recently, it was commonly accepted that, for obstetrical reasons, the development of the human brain had reached the limit of what was possible. There has been an evolutionary conflict, in our species, because a pelvis adapted to the upright posture must be narrow enough to allow the legs to be close together under the spine, which facilitates transfer of forces from legs to spine when running. (The faster our ancestors could run, the more likely they were to survive.) Hundreds of millennia ago the diameter of the foetal head at term (which is not exactly a sphere) became roughly the same size as the larger diameter of the female pelvic outlet (which is not exactly a cone), leading to the idea that the limit of foetal head size had been reached.

Since that time there have been minor fluctuations in head size. We must mention, in particular, a transitory tendency towards a reduction in head size at the time of the Neolithic revolution. Was it an effect of the 'domestication' of Homo? It is well known that among mammals such as pigs, sheep, dogs, cats, camels, ferrets and minks, one of the effects of domestication is a significant reduction in brain size.[3] The changes in the brain of a wild creature into that of a highly domesticated strain happen very rapidly in terms of evolution – after only 120 years of domestication, a brain size reduction of about 20% has been observed in mink.[4] The most plausible explanation for this is that domesticated animals have few opportunities to take the initiative, to struggle for life and to compete. Compared with wild animals, they don't have many opportunities to activate their brains. This discreet tendency towards a reduction in brain size was also probably related to the advent of agriculture and a diet based on cereal grains.

Today caesarean section has become an easy, fast, safe and therefore widespread operation, so more and more human beings are not born by the vaginal route. A tendency towards an increased head circumference at birth is therefore plausible, since the 'evolutionary bottleneck' has suddenly disappeared.[5]

Nutrition

In the early 1990s, we conducted a 'hospital and parity matched comparison study',[6] in which 499 pregnant women, attending selected clinics, were offered a 20 minute nutritional advice session before 29 weeks gestation. They were encouraged to increase their intake of oily sea fish. For each woman interviewed a corresponding control was established. The objective was to know if, in the context of a British non-teaching hospital, one short session of nutritional counselling could have measurable effects at birth. Before the 1990s, the effects of fish oil supplementation had been occasionally evaluated, but not the effects of fish consumption, and other perinatal criteria than head circumference had been taken into account.

In our preliminary study, the routine measure of head

circumference provided the only statistically significant difference between the two groups. There was a tendency towards an increased head circumference for gestational age. In another British hospital, our study was replicated and enlarged.[7] Once more, the estimated effect on mean head circumference was highly significant. There were no significant differences in terms of birth weight, length of gestation and rates of caesarean sections.

These results suggest the need for further studies about neonatal head circumference in relation to such factors as the comparative availability of land food and seafood.

Environmental pollutants

The concepts of 'timing of exposure' and 'windows of susceptibility' are considered essential in studies evaluating the possible effects of environmental pollutants. There are many reasons for focussing on foetal life, and considering the issue of brain development in particular. Once more, neonatal head circumference is of interest.

Among the neurotoxic pollutants introduced via the digestive tract is acrylamide. In 2002 we learned that acrylamide is formed in a wide variety of carbohydrate-containing foods during frying or baking at high temperatures.[8] It has since been confirmed that there are significant amounts of acrylamide in commonly consumed foods such as fried potatoes, potato chips, biscuits, breakfast cereals and coffee. The developmental toxicity of acrylamide has been confirmed by studies in rodents exposed in utero. A prospective European mother-child study examined the associations between prenatal exposure to acrylamide and birth outcomes, including head circumference.[9,10]

Development toxicity of airborne pollutants should

also inspire a generation of studies focusing on brain development. Because the placenta in our species is special (only one thin membrane between the maternal and foetal bloodstreams), particular importance should be attached to studies looking at the effects of transplacental transfer of nanoparticles smaller than 100 nanometres (such as those emitted by diesel engines). According to the State of Global Air 2017 report, inhalation of fine airborne particles is now the fifth major health risk, behind high blood pressure, smoking, high blood sugar and high cholesterol.

What we should keep in mind

To detect the most significant transformations of our species in relation to changing lifestyles, we must once more keep in mind that a huge neocortex is the main characteristic of Homo. This is why neonatal head circumference, as an available but underused index of brain weight, is an irreplaceable criterion.

6

A turning point in maternal-foetal conflicts*

Mammals in general

Among mammals in general, mother and foetus do not carry identical sets of genes: the child has maternal and paternal sets of genes. In other words, the harmony of interests between mother and foetus cannot be complete. There are reasons for conflict. To understand such conflict we must keep in mind that one of the roles of the placenta is to be the 'advocate for the baby': as an endocrine gland, the placenta can manipulate maternal physiology for foetal benefit. A maternal disease occurs when the demands of the foetus(es) via the placenta exceed what the mother can provide without creating her own imbalances.

The nature and the expression of maternal-foetal conflicts differ among different species of mammals

---

* *Author's note:* some parts of this chapter are not essential for understanding maternal-foetal conflicts, but may be of interest to those with a knowledge of biochemistry. These appear in small letters between brackets.

according to their nutritional priorities during the prenatal phase of development. For example, in some mammals, including dogs, so-called 'eclampsia' is related to hypocalcaemia (it is 'perinatal tetany'). In dogs, the priority at the end of pregnancy and the beginning of lactation is the development of the bones of the offspring, which are much more mature at birth than the bones of other mammals. Treatment is based on the intravenous administration of calcium. In ewes, veterinarians use the terms 'pregnancy disease' or 'pregnancy toxaemia'. Among these herbivorous mammals the foetus is sustained almost entirely by glucose, consuming 40% of the blood sugar produced by the mother. The disease occurs in late pregnancy. It is more common in thin ewes carrying multiple foetuses, and is characterised by a destabilisation of the glycaemia that leads to fat catabolism (this disease has also been called 'lambing ketosis'). Treatment is based on the administration of glucose.

The particular case of Homo

Interspecies comparisons encourage us to question the potential for gestational conflicts in humans, the marine chimpanzees with gigantic brains. The spectacular brain 'growth spurt' that occurs during the second half of pregnancy is a specifically human trait. A conflict between the demands of the foetus and what the mother can provide without creating her own imbalances leads us to first consider the needs of the developing brain. We present pre-eclampsia/eclampsia as a specifically human expression of maternal-foetal conflicts. This pregnancy disease, which was called toxemia in the past, is now called pre-eclampsia as long as there are no convulsions.

Several puzzling aspects of the disease are not explained by the widespread belief that faulty placentation is the central pathological process:

- The fact that there is a significant association between pre-eclampsia and high birthweight babies, in addition to the well-known association with small-for-gestational-age foetuses.[1]
- The fact that pre-eclampsia is principally a disease of first pregnancy.
- The association of pre-eclampsia with lower infant mortality in preterm babies.[2]
- The reported association of pre-eclampsia with a reduced risk of cerebral palsy.[3]
- The fact that pre-eclampsia is more frequent in the case of multiple pregnancies.
- The fact that 5 to 15 years after pre-eclampsia, women have brain alterations that cannot be explained by their cardiovascular profile.[4] This suggests that pregnant women who have difficulty meeting the specific nutritional needs of the developing brain of their unborn babies, also have difficulty meeting the nutritional needs of their *own* brain.

It is noticeable that all the well-documented metabolic imbalances that occur in pre-eclampsia are related to needs that are more easily satisfied when there is access to the seafood chain. Consider iodine, which is essential for thyroid hormone production. Thyroid hormones are needed for brain development. (In general the level of free thyroxine is low in pre-eclampsia, while the level of thyroid-stimulating hormone [TSH] is high. A low total triiodothyronine level has been reported in pre-eclamptic mothers of low birthweight babies.[5] The degree of alteration in the levels of thyroid hormones reflects the severity of the disease. Furthermore, pre-eclampsia is associated with foetal and neonatal thyroid enlargement and elevated free triiodothyronine level.[6]) In other words, the vulnerability of Homo to iodine deficiencies is reinforced during the foetal phase of development.

We have already mentioned that the huge human brain, a fatty organ, has special needs in terms of lipids, including a fatty acid (DHA) that is abundant and preformed in the seafood chain. Blood DHA concentrations remain pretty stable during pregnancy, whatever happens. It is as if, among humans, the priority is to feed the developing brain. The point is that, in the case of pre-eclampsia, a stable DHA concentration has a cost: it creates imbalances in the maternal body, particularly in the system of prostaglandins. Pre-eclampsia appears to be an essential aspect of maternal-foetal conflict in humans.

(The concept of maternal-foetal conflicts suggests we should establish a new classification of the numerous well-documented biological imbalances related to the metabolism of fatty acids.[7] It seems that the central imbalance in human pre-eclampsia is the enormous discrepancy between the maternal plasma levels of DHA – the 'brain specific fatty acid' – and EPA. In pre-eclampsia, the level of DHA is pretty stable, whereas the level of the parent molecule EPA is about 10 times lower than in normal pregnancy.[8] These are exactly the data we would expect when assuming that brain development is a priority among humans. Such data are confirmed by the Curaçao study,[9] which looked at the fatty acid composition of maternal and umbilical cord platelets from pre-eclamptic women. Whatever the circumstances, the levels of DHA remained stable. This is notable, given the difficulty humans have with making the brain specific fatty acid. The price of stable DHA is an imbalance in the omega-3 family that is at the root of a series of further imbalances. This leads to a vicious circle when the demand for long-chain fatty acids is at its greatest: if the amount of polyunsaturates available is low, the priority is to keep the level of DHA as stable as possible.

The use of biochemical markers of dietary intakes of lipids has demonstrated that a diet poor in omega-3 fatty acids is a risk factor for pre-eclampsia. Studies of the erythrocyte fatty acids profile found that women with the lowest levels of omega-3 fatty acids were 7.6 times more likely to have had their pregnancies complicated by pre-eclampsia compared to those women with the highest levels.[10] A 15% increase in the ratio of omega-3 to omega-6 was associated with a 46% reduction in the risk of pre-eclampsia. Evaluating the fatty acid composition of maternal platelets is another way to use biological markers of dietary fat intake. According to the Curaçao study, the ratio of arachidonic acid [AA – the omega-6 polyunsaturate with 20 carbons and 4 double bonds] to EPA is significantly higher in the maternal platelets of pre-eclamptic women.

These significant concordant data suggest that when the amount of omega-3 available is low, the first compensatory effect – in order to maintain an adequate supply of DHA – is a collapse in the level of EPA. This precipitating factor explains the well-known imbalances in the system of prostaglandins and particularly the decreased ratio of prostacyclin to thromboxane-2. When the level of EPA is diminished, there is no production of the physiologically inactive thromboxane-3. This leads to an overproduction of the physiologically active thromboxane-2, through a mechanism of enzymatic competition. Moreover, when the level of EPA is low, there is no production of the physiologically active prostacyclin-3.)

If we view the well-documented metabolic imbalances in this way, it appears plausible that many factors can independently increase maternal-foetal conflict. Faulty placentation, inadequate nutrition, a high birthweight baby, multiple pregnancies and certain combinations of maternal and foetal genotypes are among these possible factors. To explain why pre-eclampsia is principally a disease of first pregnancy we must recall that the metabolism of omega-3 fatty acids is influenced by parity.[11,12] The DHA content of cord blood phospholipids depends on birth order; in other words, the ability to provide pre-formed DHA is depleted with repeated pregnancies. It is as if brain development is a higher priority in the case of a first baby. The association of pre-eclampsia with lower infant mortality in preterm babies suggests that the consequence of preserving the needs of the developing foetus at any price may be a maternal disease, but the risk of infant death is reduced. The reported association of pre-eclampsia with a reduced risk of cerebral palsy can also be interpreted in this way.

Until now, studies of nutritional risk factors for pre-eclampsia have looked at the effects of nutrients, but not the effects of whole food. For example, the effects of vitamin C and E,[13] magnesium,[14] calcium[15] and zinc[16] have been explored. The relationship between vitamin D and pre-eclampsia has also been investigated.[17] Of course, there have been studies of fish oil supplementation. Most of them began during the second half of pregnancy. It is remarkable that the only study that demonstrated highly significant effects of fish oil supplementation on the risk of 'toxaemia' was conducted in London by the People's League of Health in 1938–39, when rates of severe 'toxaemia' were in the region of 6%. This controlled trial was saved from oblivion by S.F. Olsen and N.J. Secher.[18] These authors looked at thousands of pregnant women who, at random, had or had not received a dietary supplement containing vitamins, minerals and halibut liver oil from about week 20 of pregnancy. A significant effect of treatment was a more than 30% reduction in the incidence of 'toxaemia' among women expecting their first babies.

Other aspects of maternal-foetal conflicts

While pre-eclampsia is the most typical expression of maternal-foetal conflict in humans, there are others to consider. We must mention, in particular, what is commonly called 'gestational diabetes', although this term refers to the interpretation of a test rather than to a precise disease with specific symptoms and signs ('a diagnosis still looking for a disease'). When the placenta, as the advocate for the foetus, 'asks' the mother to provide a greater amount of glucose to satisfy the increasing needs of the fast-developing foetal brain, some women must make a greater effort than others. This is, more often than not, the case for women with a particular metabolic type (e.g. women fatter than average, with blood pressure higher than average, older or younger than average, or at increased risk of developing real diabetes later on in life). If such women are given a dose of sugar (as in the 'glucose tolerance test'), they have an immediate and spectacular peak of glucose in their blood. If this peak rises above a certain pre-determined threshold, the term 'gestational diabetes' is used to describe the phenomenon.

This 'diagnosis' has two practical implications. The first is that women will be told to ensure a daily amount of physical activity. The second is that they will be told to avoid high 'glycaemic index' food, such as pure sugar, soft drinks, white bread, potatoes and so on. I have suggested that the term 'gestational diabetes' should be eliminated from our vocabulary, first because it has dangerous effects on the emotional state of pregnant women: the word 'diabetes', evocative of a serious chronic disease, has the power to transform a happy woman into one who is anxious or depressed. Furthermore, it is useless: it would probably be more cost-effective to routinely spend more time discussing lifestyle in depth with *all* pregnant women.[19] Whatever the vocabulary we use, however, we must keep in mind that the difficulties some women have in satisfying the basic nutritional needs of the foetus are related to the fast development of the human brain.

Neanderthals have not disappeared

Until now, 'eclampsia', 'pre-eclampsia' and 'gestational diabetes' have been terms used by medical doctors and other health professionals involved in pregnancy and childbirth. Practical questions are raised regarding risk factors, prevention and therapeutic strategies. Viewing these conditions as part of a framework of maternal-foetal conflict should be a turning point that allows us to transcend the medical way of thinking, stimulating the curiosity of many different scientists.[20,21,22] It might even be a way to explain why some subspecies disappear while others survive.

We'll illustrate this turning point by mentioning the hypothesis offered by the French palaeontologist Jean Chaline to explain how, about 30,000 years ago, Neanderthals apparently disappeared, while their close cousins Sapiens went on to colonise the whole planet. We must recall that the average cranial capacity of Neanderthals was at least equal to the cranial capacity of modern humans. Jean Chaline suggested a possible correlation between the enormous developing brain and the risk of maternal and foetal death due to eclampsia if specific nutritional needs could not be met.[23] An enormous brain might be a factor in making human beings vulnerable in terms of survival of the species. Direct violent genocide is not the most likely explanation for the elimination of Neanderthals. All interpretations of the apparent disappearance of the subspecies Neanderthal must take into account the facts that are considered to be established:

- It took place during a comparatively short period (less than 30,000 years) of cohabitation with Sapiens. This period ended about 30,000 years ago.
- It occurred at roughly the same time in places as

diverse as the Mediterranean coast, continental Europe and Asia. This suggests that sudden food deprivation or climatic changes should not be considered probable factors.

- Sapiens and Neanderthals interbred: there were gene exchanges. Today, if we are not African, 1.5% to 4% of our nuclear DNA comes from Neanderthals.
- There is no Neanderthal DNA on the Y chromosome of modern Sapiens. This is key. It means that a human baby could be born after sexual intercourse between a male Sapiens and a female Neanderthal, but not after sexual intercourse between a male Neanderthal and a female Sapiens.[24] A probable mechanism for this is a maternal immune response (a 'histoincompatibility') during pregnancy. The expression of this response might have been a defective implantation, a miscarriage, or a pregnancy disease such as eclampsia.

Such interpretations suggest that the male Sapiens could transmit more genes than the male Neanderthal, until the time when there was no need for disparate descriptions of the two subspecies. The observable characteristics (the 'phenotype') of the Neanderthal had apparently disappeared, but not the Neanderthal genes.

The point is that scientific advances can suddenly push the limits of observable characteristics (the phenotype) of a species. This is what happened in 2014 with the advent of a new generation of studies looking at the physiological characteristics of modern humans in relation to the legacy of interbreeding between different variants of the 'big-brained Homo'.[25,26] An enormous study was based on a powerful new method for scanning the electronic health records of 28,000 Americans of European origin.[27] It appeared that some Neanderthal gene variants are associated with physiological particularities, which are in turn associated with the comparative prevalence of pathological conditions. Studies of gene expression are based on other ways to evaluate the legacy of the Neanderthals.[28]

It is now acceptable to claim that neither the genes nor the observable characteristics of Neanderthals have disappeared. Modern Homo is the result of an 'admixture' of several variants of the 'big-brained marine chimpanzee'.

7

A turning point in the attraction to water during labour

If you look at scholarly articles, textbooks for health professionals or popular books published before the 1970s, I doubt you'll find any allusion to the attraction to water during labour. Does this mean that the current widespread use of birthing pools and the publication of countless books and articles about the use of water in childbirth are just the expression of a transitory fad? Or are there reasons to think that it is the expression of a deep-rooted aspect of human nature?[1]

Before the age of tap water

We can at least affirm that legends of aquatic births and dreams associating birth and water are not new. In ancient Greece, Aphrodite, the goddess of love, was born from the foam of the waves. As for the Cyprian goddess of love, she was born on the beach at Paphos. Freud interpreted dreams of water as being about birth. According to oral tradition, Japanese women living in some small villages by the sea used to give birth in water. Engravings suggest that in some African tribes the traditional place to give birth was near a river. Some aborigines on the western coast of Australia used to paddle in the sea before giving birth on the beach. Birth under water was probably known in cultures as diverse as the Indians of Panama and, perhaps, some Maoris of New Zealand. The only valuable written document we have is a book by the American linguist Daniel Everett, who spent a great part of his life among the Pirahã, an ethnic group of Amazonians along the River Maici.[2] In the dry season, when there are beaches along the Maici, the most common form of childbirth is for the woman to go, usually alone, into the river up to her waist, then squat down so that the baby is born into the river. It is supposed to be cleaner and healthier for the baby.

Until recently, in our societies, nobody would have thought of considering the effects of a watery environment on the way women give birth. However, it is worth meditating on the widespread ritual of boiling water at births. Since this ritual preceded the Pasteur era, the rationale could not have been to make the water sterile. Occasionally, it could have been used as a way to keep a woman's husband busy. We can say for sure that it was a way to create a watery environment. It is worth noting that an interest in the issue of birth and water started to openly develop when everybody had easy access to hot and cold tap water. It is probable that until that time the attraction to water during labour could not express itself. It was simply obscured for practical reasons.

In the age of tap water

I realised the power of a watery environment in the 1970s, at a time when one of my objectives was to develop strategies to reduce the need for pharmacological assistance in the particular case of women in established labour with unbearable back pain, when the dilation of the cervix could not progress. This is why I introduced the use of intracutaneous injections of sterile water in a precise zone of the lumbar region.[3] Interestingly, in 2005 the *American Journal of Obstetrics and Gynecology* published a systematic review of 18 trials of any type of complementary and alternative therapies for labour pain.[4] All of these prospective, randomised controlled trials involved healthy pregnant women at term, and contained outcome measures of labour pain. They compared the effects of acupuncture, biofeedback, hypnosis, massage and autogenic training. Intracutaneous water injection was the only method that constantly appeared effective.

In such a context I postulated that immersion in water at the temperature of the body should also be a way to break a vicious circle, by inducing a state of relaxation and lowering the levels of adrenaline. This is why, as a temporary measure, we had a big garden paddling pool in our hospital. This was the beginning of the history of hospital birthing pools. Later on we installed a larger, round, deep blue bath that was plumbed in. On the walls of the small aquatic birthing room there were pictures of dolphins. I had a revelation the day a woman gave birth on the floor before the pool was actually full. All she needed was to *see* the blue water and *hear* the noise of water. Until that moment I had only thought of simplistic physiological explanations for the effects of water immersion during labour.

This was the first lesson: when in labour, many women are irresistibly attracted to water and great importance should be attached to the way they are introduced in the aquatic birthing room.

The second important lesson was that, in general, when a woman enters the birthing pool in well-established labour, there is spectacular progress in the dilation of the cervix during the first two hours. The progress of cervical dilation is usually associated with behaviour suggestive of a reduction in neocortical control. Ideally, the bath should be deep enough for real immersion and the water temperature comfortable (in practice around 37°C). A dim light is, of course, an important factor and, if it is acceptable, there should be nobody around apart from one silent, low-profile and motherly birth attendant who does not behave as an 'observer'. After two hours there is usually some sort of feedback mechanism and contractions become less and less efficient.

When we first used an inflatable pool (before we installed the solid pool), women were not influenced by the media or by what they had read in books about childbirth. Their behaviour was spontaneous and thus we learned about the genuine effects of a watery environment. A typical scenario (with many possible variations) was the case of a woman entering the pool in hard labour, spending an hour or two in water and then feeling the need to get out of the pool when the contractions became less effective. This return to dry land often induced a short series of irresistible and powerful contractions, so that the baby was born within a few minutes.

One day, a mother-to-be had not been in the water for long when suddenly she had two irresistible contractions and the baby was born before she felt any need to get out of the pool. While giving birth, this woman was really 'on another planet'. Clearly, in that altered state of consciousness associated with hard labour, she intuitively knew that her baby could be born safely under water. There was no panic. It is as if a deep-rooted knowing was able to express itself as soon as the intellect and its knowledge were set aside. Such births happened again.

It is as if, while in a particular state of consciousness, some women knew that a birth under water was safe for the baby.

Once we had experience of 100 babies born under water (while thousands of women had used the birthing pool), I published our observations in a mainstream medical journal.[5] This was an opportunity to warn my colleagues that in any hospital where a birthing pool is available, birth under water is bound to occur occasionally, even if it is not intentional. It is notable that my original message was distorted: many women became the prisoners of their project of 'water birth', staying in the birthing pool even when the contractions were becoming less and less effective. For that reason I tried – with limited success – to recall the original message in midwifery journals.[6,7,8]

Learning from the media

Talented journalists usually have their fingers on the pulse of what interests the general public. They can anticipate how their audience (readers, viewers or listeners) will respond emotionally. They know, or rather they have an instinct, about what will capture the public's imagination. We had a lot to learn from the incredible interest shown by journalists in birth under water. At first we were irritated by reporters who came to our maternity unit and only seemed interested in birth under water. I tried unsuccessfully to draw their attention, in the context of a general hospital, to the small home-like birthing room and the singing sessions for pregnant women. I also tried to explain how we had learned that, in humans, lactation is supposed to start during the hour following birth. I learned then the discrepancy between what is 'media-friendly' and what is reality.

In the late 1970s we were visited several times by

Jacques Mayol, the famous diver who became the hero of the film *The Big Blue*. Before we had heard about 'the aquatic ape hypothesis' through the books by Elaine Morgan, he was already convinced that Homo sapiens has special links with the sea and in particular with dolphins. One of his favourite ways to illustrate what he considered the genuine and specific relationship between humans and dolphins was to mention 'The Monkey and the Dolphin', a fable by Aesop. A ship was wrecked off the coast, close to the port of Athens. The dolphins took the shipwrecked people on their backs and swam with them to shore. When one of the dolphins saw a monkey struggling in the water, he thought it was a man and made the monkey climb up on his back. Then off he swam with him towards the shore. On the way the dolphin turned his head and realised his mistake. Without more ado, he dived, leaving the monkey to take care of itself, and swam off in search of human beings to save. This illustrates the idea that powerful compassionate links exist specifically between humans and dolphins.

Once, during a conversation with Jacques Mayol, the name of Igor Charkovsky came up: he had been associated, in the media, with birth under water in Moscow. I mentioned that I had already contacted the embassy of the Soviet Union in Paris to try to communicate with 'Dr Igor Charkovsky'. The embassy politely answered that they could not find this name in the list of Russian medical practitioners. My first thought was that this doctor was probably a legend, born in the imagination of a clever journalist. Jacques Mayol said that he had good friends in Moscow who might help him to find out the truth. Some days later I received a postcard from Moscow signed by Jacques, and Igor!

It turned out that Igor Charkovsky was not a doctor, but a swimming instructor, who had focused his interest on the capacities of human babies in water. He had realised that the younger the baby is, the better his adaptation to water. He had learned a great deal from his own daughter, who was born prematurely and, as a newborn infant, spent much of her time in water: it was supposed to reduce the need for oxygen by eliminating gravity. This is how Igor Charkovsky became involved in 'water birth'. The climax of the 'Charkovsky Phenomenon' was reached when the media reported fascinating anecdotes of babies being born in the Black Sea among dolphins!

I cannot help seeing connections between the 'Charkovsky Phenomenon' and the 'Leboyer Phenomenon': the profound effect of a succession of works of art on the general public. In both cases there is water and there are newborn babies. Of all Leboyer's numerous works, a lot of people remember only that in his book, *Birth Without Violence*, and in his first film, he gave the newborn baby a bath. Nobody knew better than Leboyer that, above all else, the newborn needs his mother's arms. But it was the artist Leboyer who gave the baby a bath. Interestingly, Leboyer introduced water into all his books and films. In the book *Loving Hands* the infant is given a bath after a massage by Shantala. At the end of the film *Loving Hands* we are on a beach, watching and hearing the waves. In the book *Le Sacre de la Naissance* the stage of labour that precedes the last contractions is symbolised by the anger of the ocean, which, instead of intimidating, stimulates. The words are reinforced by the famous painting of Hokusai, *Waves of Kanagawa*. In the silent film *Breathing* the emotions of the labouring woman are expressed through the sight and the sound of water.

Leboyer and Charkovsky have many points in common. Both of them are guided primarily by intuition, faith, beliefs, feelings, and even clairvoyance; they don't communicate their knowledge via scientific literature.

Changing the focus

Until the end of the 20th century, the dominant questions were about the short-term safety of birth under water, with a tendency to forget that the original role of the birthing pool was to avoid pharmacological assistance during labour. A study published in 1999 was a landmark in the history of birthing pools. This study was authoritative for many reasons. First, the conclusions were based on large numbers: the authors, who belonged to a prestigious department of Epidemiology and Public Health, had traced the 4,032 babies born under water in England and Wales between April 1994 and March 1996. Furthermore, several inquiries were combined in order to eliminate the effects of under-reporting.[9]

From April 1994 to April 1996, all 1,500 consultant paediatricians in the British Isles were surveyed each month by the British Paediatric Surveillance Unit and asked to report whether they knew of any births that met the case definition of 'perinatal death or admission for special care within 48 hours of birth following labour or delivery in water'. The findings were compared with reports to the confidential inquiry into stillbirths and death in infancy (a mandatory notification scheme). At the same time a postal questionnaire was sent to all National Health Service (NHS) maternity units in England and Wales, in 1995 and again in 1996, to determine the total number of births in water during the study period.

The main results can be easily summarised and remembered. There were five perinatal deaths among 4,032 births in water (a rate of 1.2 per 1,000). In the context of the UK this rate is similar for deliveries that do not take place in water. Furthermore, none of these five deaths was attributable to delivery in water: one stillbirth was diagnosed before immersion; another stillbirth occurred after a concealed pregnancy and unattended homebirth with no previous prenatal care; one baby died aged three days with neonatal herpes infection; one died aged 30 minutes with an intracranial haemorrhage after precipitate delivery, and another one, who died aged eight hours, was found to have malformations of the lungs at postmortem examination. There were 34 babies admitted for special care (a rate of 8.4 per 1,000). Rates of admission for special care of babies born to low-risk women are significantly higher than for babies born in water. Birth in water may have caused water aspiration in two babies.

Compared with well-known anecdotes, such as one case of neonatal polycythaemia (excess of red blood cells),[10] this survey of more than 4,000 babies born in water has been paradoxically ignored by the media, medical circles and the natural childbirth movement. However, it undoubtedly represents a landmark in the history of the use of water during labour. Since it was published, countless other relevant studies have been reported. The Evidence Based Birth website by Rebecca Dekker* has a wealth of information.

This new generation of research should change our focus. We must remember that the main reason to have birthing pools is to facilitate the birth process and reduce the need for drugs. We can also make a small number of simple recommendations in order to make the most effective use of birthing pools.

One such recommendation is to avoid planning a birth under water. When a woman has made such plans, she is tempted to stay in the pool while the contractions are getting weaker, with the risk, in particular, of a difficult delivery of the placenta. If all similar simple recommendations were taken into account, the use

---

* evidencebasedbirth.com/waterbirth

of water during labour would seriously compete with pharmacological assistance.

The time has now come to raise new questions about birth and water. One is about the sense of smell of newborn babies in the particular case of birth under water. Another is about the colonisation by microbes of babies born this way. As a first step, we can only evaluate the importance of these questions in the current scientific context.

8

A turning point in the microbial colonisation of newborn humans

To evaluate the possible consequences of the recent turning point in birth environments, the first step is to recall that, a century ago, most women gave birth among a great diversity of familiar microbes. Today, it is the opposite. We must also keep in mind the particularities of the human placenta.

The particularities of the human placenta

The human placenta is special because it is so effective at transferring maternal antibodies (IgG) that foetal blood concentrations of IgG are roughly the same as maternal concentrations at 38 weeks of pregnancy and continue to increase thereafter: they can occasionally reach more than twice the maternal concentrations at the time of birth.[1,2,3]

The basic needs of newborn babies must be interpreted in the light of inter-species differences regarding placental structures and functions.[4] In most mammals, which do not receive antibodies before being born, the priority is immediate access to colostrum. For them, the early colostrum is vital. Among humans, even if it is beneficial, it is not *vital*: for thousands of years, the main effect of perinatal beliefs and rituals has been, more often than not, to deprive the neonate of early colostrum. In our species, the main questions are about the bacteriological environment in the birthing place, how diverse it is and how familiar it is to the mother and therefore to the baby. Familiar microbes are friendly microbes. Today, we are in a position to understand that the millions of microorganisms that are the first to 'occupy the territory' will start programming the immune system. We must realise the importance of this topic, since it is about health development.

These considerations, inspired by immunology and bacteriology, are instrumental in evaluating the scale of the recent turning point in the history of human births. There are of course varying degrees of alteration to the bacteriological environment in the birthing place. Exposure to antibiotics and births by caesarean section in the sterile environment of an operating room are extreme examples.

When a paradigm shift pushes us to consider human births from these new perspectives, it will be impossible to avoid new questions about the future of dysregulations of the immune system, particularly the prevalence of pathological conditions such as allergic diseases and autoimmune diseases. Until now, epidemiological studies inspired by these questions have been exceptionally rare and have had to overcome technical difficulties, at a time when home birth is usually marginalised. It has been easier to contrast vaginal birth and caesarean birth inside the framework of conventional departments of obstetrics.

To address the new questions, we will need studies contrasting homebirths with vaginal births in hospital. In practice, for multiple reasons, such studies are not feasible in emerging and wealthy countries, apart from the Netherlands. A Dutch birth cohort study involving more than 1,000 children (born at a time when the rate of homebirth was above 25% in that country) included data on birth characteristics, lifestyle factors and allergic manifestations collected through repeated questionnaires from birth until age seven.[5] Faecal samples were collected at one month of age for bacteriological studies, and blood samples were collected at 1, 2, and 6 to 7 years to determine the levels of specific antibodies (IgE). Homebirth, compared with vaginal hospital birth, was associated with a low risk of allergic diseases and asthma. The differences were highly significant for children with allergic parents. Should we claim that there are two kinds of births: birth at home and birth elsewhere?

The host and the microbiome

A gigantic and highly developed brain is not the only organ that makes Homo different from other mammals. Until recently we had not realised how special the human placenta is. It is special even compared to mammals, such as primates and rodents, which share a 'haemochorial' placenta. The term 'haemochorial' means that there is only one thin membrane between the maternal blood and the foetal blood. In humans, this thin membrane includes highly active receptors for transfer of antibodies (IgG). Most mammals have several membranes separating the maternal and foetal blood circulation. There are 'epitheliochorial' placentas (horses and pigs), 'synepitheliochorial' placentas (ruminants) and 'endotheliochorial' placentas (carnivorous mammals).

In species with several placental membranes, access to early colostrum is the first opportunity for the neonate to receive maternal antibodies.

The importance of these details is easily understood in the age of the 'microbiome revolution'. Today, it is commonplace to present Homo as an ecosystem, with symbiotic interactions between the trillions of cells that are the products of our genes (the 'host') and the hundreds of trillions of microorganisms that colonise the body (the 'microbiome').

Why did it take us so long to realise the relationship between the host and the microbiome? While the host offers raw materials and shelter, hundreds of trillions of microorganisms feed and protect their host. It is as if since Pasteur's era, and until recently, there has been a deep-rooted cultural conditioning associating microbes and diseases, and classifying all microbes as enemies. This cultural conditioning was not seriously challenged while bacteriologists could only look at microscopes and cultivate microbes on Petri dishes. Things changed when those bacteriologists became geneticists, exploring the world of microorganisms by using the power of computer processing and new DNA sequencing technologies: not all bacteria seen under the microscope can be cultured, since their growth conditions are unknown. Bacteriologists can now see the 'unseen majority'.

It has been an important step in the history of science to realise that each human body is colonised by more microorganisms than there are human beings on planet Earth. This fact has inspired new questions about the multiple vital functions of the microbiome and when and how it starts being established. Until recently, it was commonplace to claim that 'to be born is to enter the world of microbes'.[6] Today we need to reconsider the premise of the 'sterile womb'. We have learned about the placental microbiome.[7] An important point is that the placental microbiome is akin to the maternal oral microbiome. The 'microbiome revolution' inspires questions that would have been considered futile until recently: an example is the function of kissing.

The function of kissing

Kissing is a universal behaviour among members of the chimpanzee family, including humans. Interestingly, there has been at least one valuable bacteriological study of 'French kissing'. Dutch researchers found that during an intimate kiss of 10 seconds there is an average total transfer of 80 million bacteria.[8] They investigated the effects of intimate kissing on the oral microbiota of 21 couples by self-administered questionnaires about their past kissing behaviour and by the evaluation of tongue and salivary samples in a controlled kissing experiment. In addition, they quantified the number of bacteria exchanged during intimate kissing by the use of marker bacteria introduced through the intake of a probiotic yoghurt drink by one of the partners prior to a second intimate kiss. Similarity indices of microbial communities show that average partners have a more similar oral microbiota composition than do unrelated individuals, with by far the most pronounced similarity for communities associated with the tongue surface.

In the framework of the microbiome revolution, we are now expecting bacteriological studies of the function of kissing in the perinatal period. Since the advent of the socialisation of childbirth, the need many women feel to kiss their baby, when they are still in a specific physiological state, has been repressed. This is a reason why it has been easier, until now, to study kissing behaviours among adults. To evaluate the importance of the topic, we must recall that not only does the placental microbiome have similarities with the maternal oral microbiome,[7] but also that the oral microbiome of infants is affected by the mode of birth.[9]

The midwife and the family dog

The new awareness induced by spectacular advances in bacteriology and immunology may have immediate practical implications, although it is premature to suggest precise 'recipes' for microbial balance. The point is to keep in mind that the body of a newborn baby has been programmed to be colonised by a great diversity of familiar microbes.

An example of the possible immediate effect of this new awareness might be the case of a homebirth midwife who has understood, probably with the help of modern physiology, that 'protection' is the keyword associated with her role. Her main objective, until now, has been to protect the labouring woman from all situations that might induce adrenaline release or stimulate neocortical activity. This midwife is now in a position to realise that her role is also to protect the newborn baby against unfamiliar microbes. She might first think about how to avoid transmitting too many of her own microbes. She will be worried by the presence in the house of unfamiliar human beings. On the other hand, she will not be worried by the family dog. The domestic dog is a carrier of familiar microbes. Furthermore, thanks to its powerful sense of smell, the dog can start identifying the newborn baby, and the odour of a human being is related to his (her) microbiome. A whole host of new topics for scientific research!

For obvious reasons, our emerging knowledge of the importance of familiar microbes cannot, overnight, have the same effects in a hospital environment. From a bacteriological perspective, there is no substitute for homebirth. However, some simple adaptive practices are possible, such as being reluctant to expose foetuses and newborn babies to antibiotics. It is possible to give more importance to skin-to-skin contact immediately after birth, and to facilitate kissing, even when the baby is born by caesarean section. It is also easy to wrap the baby in clothes recently worn by the mother and, occasionally, to put it in the arms of a person, such as the baby's father, who is cohabiting with the mother.

It is likely that various strategies will be evaluated in the near future. For example, the positive effects of the use of probiotics in the perinatal period needs to be confirmed.[10] A pilot study has been conducted in Puerto Rico of the 'gauze-in-the-vagina technique'.[11] A gauze pad is placed in the vagina in order to collect bacteria-laden secretions. Then, right after a caesarean birth, the neonate's skin and mouth are swabbed. This technique is based on the assumption that human babies have been programmed for immediate colonisation by vaginal microorganisms. There are good reasons to challenge this assumption. It is highly probable that birth with intact membranes ('birth in the caul') was frequent before the socialisation of childbirth. It is still comparatively common in modern women who give birth in an environment compatible with a 'foetus ejection reflex'.[12] It is notable that many societies have observed that human beings born in the caul (i.e. protected against vaginal microbes) will be healthy (and lucky).

To compensate the effects of microbial deprivation in the neonatal period, I have suggested the use of 'Bacille de Calmette et Guérin' (BCG).[13,14] As a way to provide non-virulent microbes of the mycobacteria family, BCG is an immunomodulator previously used as a vaccine against tuberculosis and as a therapeutic agent for a great diversity of diseases. Theoretically, it should facilitate the kind of deviation of the immune system that is hindered in a more or less aseptic environment. It is the only infancy vaccine that has been evaluated through randomised controlled trials with long-term follow-up (in one study the follow-up was 60 years).[15] It is notable that the long-term non-specific effects on health were always positive. Trials with follow-up periods of less than 10 years would easily measure the prevalence of some of the dysregulations of the immune system that have become more frequent during the past decades.

There will probably come a time when it is impossible to ignore the bacteriologic and immunologic perspectives when addressing the issue of homebirth versus hospital birth.

An effect of such a historical turning point should be to 'de-marginalise' homebirth and therefore make it safer. Questions should be raised about how to adapt homebirth to the dominant urbanised lifestyle. Can a new awareness lead to a situation in which hospital staff are at the service of an increased number of women who prefer to keep the option of homebirth open? We will need to overcome our deep-rooted cultural conditioning and get out of the 'helping-guiding-controlling-supporting-coaching-managing' paradigm. We will need to understand the concept of the 'protection of an involuntary process'.

9

A turning point in the classification of human births

Reasons for offering multiple classifications of human births have only recently emerged. There was, however, a preliminary turning point during the 18th century, when the man-midwife advanced from being an attendant for the emergencies of childbirth, particularly thanks to the use of forceps, to gaining a hold on midwifery.[1] This was the time when 'lying-in hospitals' began to be founded, in part because medical men sought a ready source of clinical material for their own and their pupils' study. The oldest charitable institutions were the Lying-in Hospital in Dublin and the Hotel-Dieu in Paris. Although most women continued to give birth in their local environment, it was possible to distinguish between homebirth and institutionalised birth.

Today there are countless ways to classify the modes of birth. We have already suggested the bacteriological perspective. We have also contrasted birth on dry land and birth in water.

Pre-labour caesareans

In the age of easy, fast and safe caesarean, we should first contrast birth without labour and all the other ways to be born. Today more and more babies are born by caesarean section performed before labour starts. At a time when it appears that stress hormones have multiple roles to play, and when the concept of 'stress deprivation' has emerged in scholarly articles, we can consider birth by pre-labour caesarean as a unique, extreme and unprecedented example of stress deprivation.

It has been understood for several decades that 'pre-labour caesarean' is a risk factor for respiratory difficulties in the hours and days following birth, and that the risks are dependent on gestational age: differences in the quality of respiratory function are detectable when comparing pre-labour births at 38 and 39 weeks.[2] One interpretation of this is that the foetus participates in the initiation of labour, probably through the release – when the lungs have reached a certain degree of maturity – of specific factors (particularly surfactants).[3,4] Furthermore, the roles of maternal and foetal stress hormones are well known. The effects of maternal corticosteroids on foetal lung maturation have practical implications, and labour implies the action of beta-endorphins (releasers of prolactin, which facilitate respiratory function).[5] It also implies the release of the main foetal stress hormone (noradrenaline), which is probably an essential factor in lung maturation.

The multiple negative effects of stress deprivation among babies born by pre-labour caesarean have been underestimated until recently. For example, we now know that, under the effect of noradrenaline, the sense of smell has reached a high degree of maturity at birth among babies born by 'in-labour' caesarean or via the vaginal route. A Swedish experiment exposed babies to an odour for 30 minutes shortly after birth and then tested their for their response to this odour (and also to another odour) at the age of three or four days.[6] Since the concentration of noradrenaline had been evaluated, it was easy to conclude that foetal noradrenaline released during labour is involved in the maturation of the sense of smell. The paramount role of the sense of smell immediately after birth is clear. In the 1970s I already mentioned that the sense of smell is the main guide the baby has to find the nipple during the hour following birth.[7,8] It has also been shown that it is mostly through the sense of smell that the newborn baby can identify its mother (and, to a certain extent, that the mother can identify her baby).

Among other significant studies, we must mention one that evaluated adiponectin concentration – an agent involved in fat metabolism – in the cord blood of healthy babies born at term. The concentration is significantly lower after pre-labour caesarean compared with in-labour caesarean or vaginal birth.[9] These data suggest a mechanism according to which stress deprivation at birth might be a risk factor for obesity in childhood and adulthood. Other investigations have measured the concentration of melatonin in cord blood. It is low after pre-labour births.[10] This is an important point, since melatonin has protective anti-oxidative properties. Furthermore, these studies confirm that the 'darkness hormone' is involved in the birth process. Further work is needed to understand the role of melatonin during labour, as we are learning about a synergy between its uterine receptors and oxytocin receptors.

In general, a baby born after a pre-labour caesarean is physiologically different from other babies. For example, babies born by pre-labour caesarean tend to have a lower body temperature during the first 90 minutes following birth.[11]

Other effects of pre-labour caesarean will probably be studied in the near future. It seems that the prevalence of placenta praevia is significantly increased only in the case of a pregnancy following a pre-labour caesarean.[12]

All aspects of the period surrounding birth must be considered when analysing the particularities of 'birth without labour'. There is already an accumulation of data confirming the negative effect of pre-labour caesarean on breastfeeding, particularly at the phase of initiation of lactation.[13,14]

We must also give great importance to data describing the milk microbiome. There are significant differences in the milk of mothers who give birth by pre-labour caesarean and those who give birth by in-labour caesarean or the vaginal route.[15] The data suggest that there are other factors than the operation *per se* that can alter the process of microbial transmission to milk. Similar differences were found by a Canadian study of the gut flora of four-month-old babies.[16] Joanna Holbrook and her team, in Singapore, suggest interpretations for these surprising data after collecting faecal samples of babies until the age of 18 months. It appears that, apart from the route of birth and exposure to antibiotics, a shortened duration of pregnancy tends to delay the maturation of the gut flora: one week more or less in the duration of pregnancy is associated with significant differences: a pre-labour caesarean implies the association of all the known factors that can delay the maturation of the gut flora. This study is all the more important since it reveals that a delayed maturation of the gut flora is a risk factor for increased adiposity at the age of 18 months.[17]

In spite of possible inter-species differences, we must seriously consider animal experiments suggesting that the stress of labour influences brain development. Such is the case of studies demonstrating that the birth process in mice triggers the expression of a protein (uncoupled protein 2) that is important for the development of the hippocampus.[18] In humans, the hippocampus is a major component of the limbic system. It has been compared to the conductor of an orchestra, directing brain activity. It has also been presented as a kind of 'physiological GPS system', helping us navigate while also storing memories in space and time: the work of three scientists who studied this important function of the hippocampus was recognised by the award of the 2014 Nobel Prize in physiology and medicine. This is also the case of studies with rats suggesting that oxytocin-induced uterine contractions reverse the effects of the important neurotransmitter GABA: this primary excitatory neurotransmitter becomes inhibitory.[19] If uterine contractions affect the neurotransmitter systems of rats during an important phase of brain development, why would the same not occur in humans?

Births with labour

All births that are not pre-labour caesareans are 'with labour', whatever the route. We must first keep in mind that caesareans performed in real situations of emergency are associated with comparatively poor short-term outcomes. This well-known fact is easily interpreted. Such caesareans are often performed when there are already signs of foetal distress, after a long period of pharmacological assistance. We must also take into account that emergency caesareans are often performed in a hurry and therefore poor technical conditions. Furthermore, they are associated with an increased risk of negative long-term outcomes. For example, according to an American study, women with a full-term second stage caesarean have a spectacularly increased rate of subsequent premature birth (13.5%), compared to women who had a first-stage caesarean (2.3%) and the overall national rate (7–8%).[20]

Our overview of the multiple effects of pre-labour caesarean, associated with a reminder of the particularities of last-minute emergency caesarean, suggests that the 'ideal' caesarean is one performed during labour, before a real emergency.

Until now, 'planned in-labour caesarean' and 'in-labour non-emergency caesarean' have not been examined in epidemiological studies. In the well-known multi-centre randomised controlled trial about breech presentation at term, only two options were considered: planned pre-labour caesarean and planned vaginal birth.[21]

Once the concept of an 'in-labour non-emergency caesarean' becomes familiar, the doors will be open for a simplified strategy of overseeing birth, with two basic scenarios: either the birth process is straightforward and occurs vaginally, or it appears difficult and an in-labour caesarean before an emergency occurs is the best option. But before such a simplified strategy can become possible, midwifery and obstetrics will have to change in several ways.

The first, and most important step will be to challenge the effects of thousands of years of tradition and cultural conditioning. This is becoming possible in the light of the concept of neocortical inhibition. The key will be to study how some human physiological functions – such as the birth process – are obscured by the activity of a powerful neocortex, and to understand the solution nature found to adapt to human particularities.

Finally, we are reaching a phase in the history of midwifery and obstetrical practice when an in-labour non-emergency caesarean may in many cases be the best alternative to drug-free childbirth. In this context, we will need a new generation of tests in order to decide early enough during labour that the vaginal route is acceptable, without waiting for the phase of real emergency.

The 'birthing pool test' is an example of a tool that can be adapted to futuristic strategies. It is based on a simple fact that is worth mentioning again. When a woman in hard labour enters the birthing pool and is immersed in water at body temperature, spectacular progress in cervical dilation usually occurs within an hour or two. If well-advanced dilation does not progress despite water immersion, privacy (no camera!) and dim light, one can conclude that there is a major obstacle. There is no reason for procrastination. It is wiser to perform an in-labour non-emergency caesarean immediately.

In the 1970s and early 1980s, I had several opportunities to describe why we originally introduced birthing pools in a French state hospital. I also described the most typical scenario:

> 'We tend to reserve the pool for women who are experiencing especially painful contractions (lumbar pain, in particular), and where the dilation of the cervix is not progressing beyond about 5cm. In these circumstances, there is commonly a strong demand for drugs. In most cases, the cervix becomes fully dilated within 1 or 2 hours of immersion…'[22]

At that time, I could only refer to 'most cases'. Later, I analysed outcomes in the rare cases when dilation had not progressed after an hour or two in the bath. I realised that in the end a caesarean had always been necessary, more often than not after long and difficult first and second stages. This is how I started to tacitly take into account what I had not yet called the 'birthing pool test'.

More recently I mentioned the birthing pool test during information sessions for doulas. They reported the circumstances of many births in London hospitals, which helped me to learn. It is obvious that many long and difficult labours, with the usual range of drugs preceding an emergency caesarean, would be avoided if the birthing pool test had been used and understood. One of these anecdotes is particularly significant. A woman in hard labour arrived in a maternity unit with her doula with the dilation of the cervix already well advanced. Soon after, she entered the birthing pool. More than an hour later, dilation had not progressed. The doula, who was aware of the birthing pool test, was adamant that this woman could not safely give birth by the vaginal route. A senior doctor was eventually called and diagnosed a brow presentation. A brow presentation is difficult to diagnose in early labour, and is incompatible with vaginal birth. In this case, the doula knew that a caesarean would be necessary, although she could not explain why.

The birthing pool test implies that an internal exam has been performed just before immersion so that, if necessary, a comparison will become possible after an hour or two. This is an important practical detail, because midwives who are familiar with undisturbed and unguided births in silence, semi-darkness and privacy can usually follow the progress of labour using other criteria than a repeated evaluation of the dilation of the cervix.[23]

Today, we can offer a physiological explanation for why immersion in body-temperature warm water makes contractions more effective for a limited period of time. When a woman enters the pool in hard labour, there is immediate pain relief, and therefore an immediate reduction in the levels of stress hormones. Since stress hormones and oxytocin are antagonistic, the main short-term response is usually a peak of oxytocin and therefore spectacular progress in dilation.

Later, there is a long-term complex response, which is a redistribution of blood volume. This is the standard response to any sort of water immersion. There is more blood in the chest.[24] When the chest blood volume is increased, certain specialised cells in the atria release a peptide commonly called ANP (atrial natriuretic peptide) that interferes with the activity of the posterior pituitary gland.[25] We can all observe the effects of the reduced activity of our posterior pituitary gland after being in a bath for a while: we pass more urine. This means that the release of vasopressin – a water-retention hormone – is reduced. In fact, the chain of events is not yet completely understood.[26] We have recently learnt that oxytocin – the love hormone – has receptors in the heart (!) and that it is a regulator of ANP.[27]

In practice, we just need to remember that the immediate peak of oxytocin following immersion in warm water will induce a feedback mechanism and that the uterine contractions will become less effective after an hour or two.

In the near future, it may become relevant to make the classification of human births still more complex. We are expecting, in particular, a generation of studies evaluating the possible long-term effects on the child of different components of modern pharmacological aids to birth. What if it becomes clear, for example, that to be born by an in-labour non-emergency caesarean section is better, in the long-term, than to be born after the mother has spent several hours on an oxytocin drip?

10

A turning point towards the symbiotic revolution

At a time when the limits of the domination of nature have been reached, there are several ways to formulate the main questions that are relevant for humanity. In the late 1970s, when wondering how respect for Mother Earth – as a facet of love – develops, I had mostly in mind what is now called 'emotional intelligence'.[1] At the end of the 1990s, when rephrasing this question in the emerging framework of 'The Scientification of Love', I had mostly in mind what we are learning from fast-developing scientific disciplines.[2] And at the end of the second decade of the 21st century, I have mostly in mind what have been the dominant ways of thinking for thousands of years. It is common to present the Neolithic revolution as the advent of agriculture, animal husbandry and a sedentary lifestyle. But it was also the advent of new ways of thinking. Today we have to phrase questions in terms of paradigm shift.

The roots of the dominant ways of thinking

Before wondering if a paradigm shift is possible, we must first analyse the roots of the current dominant ways of thinking, with the help of well-known written documents. One of them is the Bible. It may be thought of as a library of texts written over the centuries that provide valuable information about the main characteristics of post-Neolithic lifestyles. It was probably preceded by an oral tradition. The bases of our dominant cultural conditioning are perfectly summarised:

> '...Be fruitful, and multiply, and replenish the earth, and subdue it; and have dominion over the fish of the sea, and over the fowl of the air, and over the cattle, and over all the earth, and over every living thing that moveth upon the earth'.[3]

> 'Thou shalt not covet thy neighbour's house, thou shalt not covet thy neighbour's wife, nor his manservant, nor his maidservant, nor his ox, nor his ass...'[4]

It is obviously the mission of the male adult human to dominate all aspects of nature, including children: 'Foolishness is bound up in the heart of a child; but the rod of correction shall drive it far from him'.[5]

It is highly probable that the domestication of animals was, to a great extent, at the root of 'the rule of the father'. With the advent of animal husbandry, the procreative power of males became obvious. The man was the one bringing the 'seeds'. The role of the woman was to be an incubator and provider of food. Before that time the procreative role of males was misunderstood, underestimated and even ignored. This was probably related to the long interval in our species between sexual intercourse and childbirth. The phenomenon of

'Venus figurines' is symbolic of the mystery of life. Over a hundred such figurines have been found all over the world, the oldest one dated to more than 35,000 years ago. All these figurines exaggerate the parts of the body related to pregnancy, childbirth and breastfeeding, particularly the hips, abdomen, breasts, thighs and the vulva: the mystery of life was obviously related to the female principle in general.

Interpreting scientific advances

In the current scientific, technological and conceptual context, the bases of our dominant ways of thinking are easily perceptible. A typical example is the way we interpret and transmit scientific advances.

From the work of Darwin we remember the concepts of the 'struggle for life' and the 'survival of the fittest'. The dominant way of thinking has stifled the voice of those who, since the 19th century, have presented 'mutual aid' as a chief factor in evolution. In his book, originally published in 1900, Peter Kropotkin completed his own observations by presenting an impressive review of available documents focusing on this factor.[6] Kropotkin attached great importance to an address given in 1880 by Professor Kessler, Dean of St Petersburg University, at a congress of Russian naturalists. Kessler did not deny the struggle for existence, but he was 'inclined to think that in the evolution of the organic world – in the progressive modification of organic beings – mutual support among individuals plays a much more important part than their mutual struggle'. At the end of the 20th century this alternative way of thinking has been eloquently summarised by Dorion Sagan and Lynn Margulis: 'Life did not take over the globe by combat, but by networking'.[7]

We might make similar comments by recalling how the

work of Pasteur has been perceived. It has been culturally easy to accept that enemies had been identified in the struggle for life. It became clear that we had to learn to dominate the world of microorganisms… an emerging aspect of the domination of nature. The vocabulary initiated by this new scientific era is highly significant: *a-sepsis, anti-sepsis, anti-biotics*, etc. Once more, the deep-rooted way of thinking has stifled the voice of those who offered another vision of the relationship between Homo and microorganisms. This is the case of the physiologist Claude Bernard, who placed great importance on '*Le Terrain*' (the figurative sense of '*le terrain*' is the basic condition, the temperament, the ability to cope with disease). It is also true of Antoine Béchamp, who understood the mechanism of fermentation and knew about microorganisms ('microzymas') before Pasteur.[8,9] He dared to say: 'Instead of trying to determine what abnormal conditions are composed of, let us first know the normal conditions which make us healthy'. Such points of view were expressed again at the beginning of the 20th century, particularly by the physiologist Jules Tissot.

Today, at a time when the mechanisms for transformations of species must be reconsidered, and in the age of the 'microbiome revolution', all these scientists, who had been running against the current, may suddenly be classified as pioneers. They have participated in the initiation of a possible turning point towards the 'symbiotic revolution'.

To understand this term, we must first realise that many aspects of the domination of nature have developed at such high speed that we might describe the current situation using phrases such as 'being at the edge of the precipice'. This analogy clearly indicates that, to survive, humanity urgently needs to take another direction. To qualify the counter-revolution that might prolong the survival of mankind, we need a term that would be the antithesis of 'domination' (of nature). The best term I can suggest is 'symbiosis', since it literally means 'with' life. This term was coined in 1879 by the German surgeon and botanist De Bary, who described lichens as the result of an association between a fungus and an alga.

A multifaceted counter-revolution

A symbiotic revolution would have several main aspects. One would be an increased human capacity for networking with other living creatures, particularly microorganisms, which are the foundations of all ecosystems. Another would be to improve our understanding of the laws of nature in order to work with them instead of neutralising them: in practice, this means training ourselves to think like physiologists, who study what is universal and cross-cultural.

Nobody would dare to describe the multiple effects of such a hypothetical counter-revolution. However, we have good reasons to focus on childbirth, as a critical phase in human development highly influenced by cultural conditioning. Since such a turning point implies new ways of thinking and therefore a renewed language, one of the necessary steps would be to prohibit disempowering terms suggestive of domination by the cultural milieu ('the patient', 'labour management', 'coaching', 'supporting', etc.) and to use active verbs such as 'giving birth'. The physiological perspective can help us to accept that one can only 'protect' an involuntary process.

11

A turning point in the history of orgasmic states

The Neolithic phase in the history of mankind is usually presented as the advent of agriculture, animal husbandry and a more sedentary lifestyle. To justify the frequent association of the terms 'Neolithic' with 'revolution', we must keep in mind many other aspects of this 'turning point'.

Since the 'marine chimpanzee' is characterised by a gigantic brain, we must first stress that it is at precisely the time of the 'Neolithic revolution' that the human brain started to decrease in size: studies of skeletal samples from different regions of the world are consistent in confirming significant reductions in skull capacity. Although several theories have been put forward, it is most likely that with agriculture came a monotonous diet based on cereal grains, with the addition of dairy food.

Regulated human physiological functions

It is also at the time of the 'Neolithic revolution' that

human physiological processes related to reproduction started to be organised, controlled, regulated and therefore repressed by the cultural milieu. This essential aspect of the domination of nature must be looked at in the current scientific context.

Today the different episodes of human reproductive life are better understood if studied together. When considering sexual intercourse, childbirth and breastfeeding, the same basic hormonal cocktails are involved, the same 'physiological brakes' may be activated (neocortical inhibition and hormones of the adrenaline family), and similar scenarios are reproduced. There is always a rather passive first phase, followed by an 'ejection reflex': sperm ejection reflex, foetus ejection reflex, milk ejection reflex… Furthermore, there are similarities in terms of subjective experience.

Genital sexuality

Societies have developed and transmitted from generation to generation many subtle ways to control and restrain all aspects of genital sexuality.

The choice of mates, in other words the organisation of marriages, is the most universal. There is a great diversity in the details of matrimonial arrangements. The wife is often considered the property of the husband. In many societies men obtain wives only through difficult or extreme means, namely bride-price, bride-wealth, or through the exchange of a female relative. Marriage rules may favour locating the new couple near the male kin of either spouse. Whatever the variants, the concept of marriage rules is an aspect of the Neolithic revolution. Daniel Everett, an American linguist who spent several decades among an Amazonian pre-Neolithic ethnic group, simply noted that 'Sex and marriage involve no ritual'.[1]

Countless aspects of universal sexual repression have been documented. Presenting sexual renunciation as a virtue is a subtle one. However, the most extreme aspect of this imposed renunciation of sexual pleasure is undoubtedly female genital mutilation. Female circumcision (when only the hood of the clitoris is supposed to be removed), clitoridectomy (when the whole clitoris is removed) and infibulation (when labia majora and minora are also cut away) are specifically and consciously designed to reduce a woman's capacity to obtain sexual pleasure. In ethnic groups where these rituals prevail, a girl who attempts to avoid the operation is socially ostracised. Uncircumcised women are considered 'unclean'. To call a woman 'the uncircumcised one', or a man 'the son of an uncircumcised woman', is a gross insult.[2] Interestingly though, despite these large-scale human experiments in reproduction, even after clitoridectomy or infibulation some women are still able to have strong sexual feelings.[3]

Male genital mutilation is usually less invasive than female mutilation, but is more widespread and has been practised on all five continents, even in places such as Oceania, the Asian and Pacific islands and the New World. Incision usually consists of either a simple cut to the foreskin to draw blood, or a complete cutting through the foreskin in a single place to partly expose the glans. In circumcision itself, the foreskin is completely removed. Subincision, which is usually associated with circumcision, was practised in particular among Australian aborigines; it involves cutting open the urethra on the underside of the penis down almost as far as the scrotum. Castration, another form of genital mutilation, was most often forced upon young boys who, as captured slaves, would supply the 'needs' of the Arab and Turk system. The use of castration as a punishment, by both individuals and by the state, has been institutionalised in many parts of the world, particularly in the Near East.[4]

Through the practice of compulsory genital mutilations, cultural milieus send significant, although discreet, messages to members of their group. One of the messages is that genitalia belong to the community rather than to individuals, which means that their use must follow strict established rules. Another conditioning effect of such rituals is that the genital parts of the body may remain associated with pain rather than pleasure.

This aspect of the domination of nature also includes topics such as taboos related to virginity and masturbation.

Childbirth

The cultural control of childbirth is another aspect of the domination of nature that started with the Neolithic revolution. We have a sufficient amount of data at our disposal to claim that in pre-agricultural societies women used to isolate themselves to give birth. The cultural control of childbirth has been direct and indirect. It has been direct with the advent of midwifery and other basic aspects of the socialisation of the event. It has been indirect with the transmission from generation to generation of deep-rooted beliefs and rituals, and the development of a powerful and widespread cultural conditioning. According to this dominant cultural conditioning, a woman is supposed to 'be delivered' by a 'helper'. Over the millennia many variants of the socialisation of childbirth have appeared and disappeared. Many institutions have been involved, particularly the religious ones. Today, with the masculinisation and the medicalisation of the event, the ultimate forms of socialisation have been reached, at such a point that childbirth is the first basic human physiological function that has been made redundant at the scale of the species: pharmacological and surgical substitutes are available and widely used.

Lactation

Since there is a continuum between birth physiology and the physiology of lactation, and since cultural milieus regulate genital sexuality and therefore the expression of libido, which is modified by lactation hormones, we cannot ignore questions about the control of infant feeding. In fact, for thousands of years, all cultures have actively interfered with lactation, at least with its initiation. I have suggested that colostrum might be regarded as a symbol of the repression of instinctive forces, since babies in many cultures are deprived of it. The quasi-universal belief that colostrum is tainted or harmful is one of the countless ways in which cultures meddle with the newborn baby's relationship with his or her mother.[5] Delaying the initiation of breastfeeding has been the rule in almost all societies we know about, including in Western Europe. In Tudor and Stuart England, colostrum was openly regarded as a harmful substance, to be discarded. The mother was not considered 'clean' after childbirth until the bloody discharge called 'lochia' had stopped flowing. She was not permitted to give the breast until after a religious service of purification and thanksgiving called 'churching'. Meanwhile the baby was given a purgative made from such things as butter, honey and sugar, oil of sweet almonds or sugared wine. Paintings from that time show the newborn infant fed with a spoon while the mother recovered in bed. In Brittany the baby was not put to the breast before baptism, which took place at the age of two or three days. The Bretons of old believed that if the baby swallowed milk before the ceremony, the devil might enter the baby's body along with the milk.

The duration of breastfeeding is undoubtedly influenced by family structures. Since human societies organise mating and create marriage rules, they also indirectly influence the duration of breastfeeding, to the extent that nobody knows exactly what the physiological ideal for the duration of breastfeeding is in humans. For any other mammal, the answer is simple – almost as simple as the duration of pregnancy.

A turning point in our understanding of the 'highways to transcendence'

In the light of recent scientific advances, it is becoming easy to interpret the subjective experiences that may culminate during 'ejection reflexes'. The existence and properties of complex 'hormonal cocktails' involving physiological agents such as natural opiates and oxytocin have recently been revealed. They appear, theoretically, as the bases of emotional states we can regard as ecstatic. We are reaching a time when the universal human transcendent emotional states are entering the framework of acceptable and even serious scientific topics.

It is notable that, in spite of deep-rooted cultural conditioning, there have always been marginalised voices associating 'ejection reflexes' with transcendent emotional states. One example from Western culture is Helen Deutsch – the first woman doctor to graduate from the University of Vienna and a member of the Vienna Psychoanalytical Society. Having experienced birth and breastfeeding, she considered sexual intercourse and giving birth as two phases of one process divided only by a time interval: 'Just as the first act contains an element of the second, so the second is impregnated with pleasure mechanisms of the first. I even believe that the act of birth represents the acme of sexual pleasure...' Furthermore, according to her, breastfeeding is 'an act of sexual enjoyment, at the heart of which the mammary gland plays the part of an erogenous zone'.[6] Una Kroll is another who has eloquently highlighted the links between diverse ecstatic states. Having been a nun, a medical doctor, a priestess and the mother of four children, she could authoritatively write: 'moments of ecstasy have recurred like grace notes throughout my life...The ecstasy of sexual union is akin to that of ecstatic prayer...'[7]

In Eastern cultures, there has been the age of female Tantric masters. A parable found in a Tantric text is highly significant.[8] It is the story of a hermit pilgrim in search of 'The Supreme Truth'. He had been travelling, meditating, fasting and inflicting unbearable pain upon himself for many years, but he felt he could never reach the Supreme Truth. One day, disillusioned by years of unrewarded effort, he rested in the late afternoon by a river. A female Tantric Master came along, intending to bathe and anoint her body. After listening to the pilgrim's story, she seduced him by 'carrying his senses through Tantric pleasures to the state of extreme arousal, wherein he found the centre of power he sought, awaiting him in what he had so long denied himself.'

There are extreme contradictions between these voices, which have been more or less neutralised, stifled and forgotten, and the dominant post-Neolithic cultural conditioning.

After thousands of years of powerful control of physiological processes by cultural milieus, it is not surprising that simple natural paths to transcendence have been ignored or stifled. Mystical emotions have often been associated, on the contrary, with sexual renunciation, celibacy, and virginity, while orgasmic states have been associated with culpability.[9] The best known and most influential mystics did not consider the transient ecstatic states of orgasms as possible paths to permanent transformation of consciousness, be they male mystics such as Shankara, Ibn Arabi or Meister Eckhart, or even female mystics such as Hildegard of Bingen, Julian of Norwich, Catherine of Siena or Clare of Assisi. It is as if many aspects of human nature had been concealed for thousands of years.

Advances in physiological sciences suggest a simple way to summarise the current situation. One can claim that during the 'orgasmophobic' phase of the history of mankind cultural milieus have regulated and controlled access to another reality beyond space and time. They have channelled the human capacity for transcendence by favouring routes that can be easily organised and regulated, such as prayer, fasting, ritualised music, songs, dances and states of trance. They have ignored, denied, stifled or made morally unacceptable the natural routes. Let us introduce, in modern scientific language, an ironic note, by noticing that cultural milieus have denied access to transcendence through opiates released by the human body, while they made acceptable pretty similar emotional states if reached through the use of psychedelic drugs.

What a turning point in the domination of nature it will be on the day when simple routes to transcendence are made culturally acceptable!

12

A turning point in our relationship with time

It is widely accepted that the long-lived radioactive wastes, including spent nuclear fuel, must be contained and isolated from humans and the environment for periods ranging from tens of thousands to a million years. Furthermore, even in the case of a storage space in granite hundreds of metres below the ground, agencies preparing for long-term waste repositories in northern latitudes must take into account possible future glaciations creating internal strains. This is one way, among others, to refer to the possible long-term consequences of the actions of modern humans. It is also a way to illustrate the turning point beyond which exploring the far future becomes imperative.

Homo was not originally programmed to think long-term. Before the Neolithic revolution, our ancestors were finding their food on a day-to-day basis. Their awareness of the process of ageing had no practical implications. The advent of agriculture and animal husbandry was the first turning point in the relationship between our ancestors and time. They were obliged to think at least in terms of seasons.

This reference to the past, and in particular to the previous crisis in the history of mankind, explains why modern humans must overcome deep-rooted difficulties to enlarge the concept of time and, in particular, to develop their capacity to explore the future.

A training tool

It is in such a context that we present our database (primalhealthresearch.com) as a tool to train ourselves to think long term. It is an imperative training, particularly among those involved in pregnancy and childbirth. There are deep-rooted and understandable reasons why most midwives, obstetricians and birth activists cannot easily see beyond the perinatal period. Traditionally everybody was happy when mother and baby were alive and healthy after what is often considered the most dangerous phase of human life. The point is that there has not been any significant paradigm shift until now. The usual modern criteria to evaluate the practices of obstetrics and midwifery are still short term. In medical language, they are called perinatal and maternal mortality and morbidity rates. In endless discussions about the comparative safety of hospital births and homebirths, both medics and 'natural childbirth' groups focus on these conventional criteria.

Our database includes published epidemiological studies that explore correlations between what happens during the 'primal period' and what happens later on in terms of health and personality traits. The primal period includes foetal life, the period surrounding birth and the year following birth. One of the objectives is to provide valuable information to those who have a special interest in the early development of human beings.

Just sifting through the database is a way to enhance curiosity and to deviate from the dominant ways of thinking. Let us imagine, for example, a midwife exploring our database. By chance, she comes across the keyword 'prostate cancer'. Although 'prostate cancer' is not usually a major preoccupation in midwifery circles, she is curious enough to click on this keyword. She is surprised and fascinated by her findings. She will first look at a Swedish study published in an authoritative medical journal in 1996 (an early phase in the history of primal health research). It is a study involving hundreds of men born at Uppsala University Hospital between 1874 and 1946. It appeared that men born to pre-eclamptic mothers had a negligible risk of developing prostate cancer in the future. Then the curious midwife will find the results of another Swedish study revealing that high birthweight is associated with an increased risk of prostate cancer, while another study revealed that prolonged pregnancy is associated with a reduced risk. The midwife will finish her enquiries by looking at a Norwegian study that considered the risks of metastatic forms of prostate cancer in relation to birth size.

We can imagine the increased curiosity of this midwife and the many questions that may cross her mind. Why have such studies never been replicated outside Scandinavian countries? There is an easy answer: in Scandinavian countries there is a long tradition of keeping detailed medical files that facilitate epidemiological studies. Why are the results of such studies not better known? Once more we can suggest an easy answer: in the age of ultra-specialisation, those interested in pre-eclampsia, birth weight and duration of pregnancy are not interested in prostate cancer... and vice versa.

The concept of timing

There are other reasons why explorations of our database can influence our relationship with time. Most non-communicable diseases are not purely genetic. They are the effects of an interaction between genetic and environmental factors. From a practical perspective, it is not essential to identify the genes involved in particular diseases. Nor is it essential to quantify the comparative roles of genetic and environmental factors. The main practical questions are in terms of timing. Our database has become a unique tool that provides some clues about critical periods for gene-environment interaction. From an exploration of the whole database, it appears that often the nature of an environmental factor is less important than the period of exposure to this factor. To illustrate the importance of the concept of timing we'll compare the data provided by the productive keywords 'schizophrenia' and 'autism'.

An overview of the studies reached through the keyword 'schizophrenia' will convince anyone that, in the case of this particular condition, the main critical period for gene-environment interaction takes place during foetal life, long before the onset of the symptoms (in early adulthood or late adolescence). For example, there are concordant results from Holland and China confirming that prenatal exposure to famine is a risk factor. There are also concordant results about the effects of extreme maternal emotional states during pregnancy after events such as natural disasters (e.g. the 1953 Dutch flood disaster) or the death of a close relative. Many studies complement one another. For example, studies of schizophrenia following prenatal exposure to influenza epidemics help to interpret studies that reveal increased risks of schizophrenia among those born in late winter or spring. We must keep in mind that schizophrenia is often associated with physical particularities (such as abnormalities of the mouth or asymmetry of fingertips) that are related to events occurring in the middle of foetal life. We must also mention studies about the risks of schizophrenia in relation to maternal exposure to toxoplasmosis, and also in relation to the use of pharmacological treatments in pregnancy (diuretics and analgesics). Significant negative findings are also important to identify the critical period for gene-environment interaction: there is no evidence of any effect of the mode of birth or the mode of infant feeding.

The keyword 'autism' offers another good example of concordant results provided by a great number of studies about a disease with a strong genetic component. From an overview of all these studies it appears that, in the case of autism, it is the period around birth that appears critical for gene-environment interaction.

My interest in autism started in 1982, before the age of modern epidemiology, when I met Niko Tinbergen, one of the founders of ethology, who shared the Nobel Prize with Konrad Lorenz and Karl Von Frisch in 1973. As an ethologist familiar with the observation of animal behaviour, he studied in particular the non-verbal behaviour of autistic children. As a 'field ethologist' he studied the children in their home environment. Not only could he offer detailed descriptions of his observations, but he could also list factors that predispose someone to autism or exaggerate the symptoms.

Tinbergen found such factors evident in the period surrounding birth: induction of labour, 'deep forceps' delivery, birth under anaesthesia and resuscitation at birth. Interestingly, this pioneer introduced the variable 'labour induction'. When I met him he was exploring possible links between difficulty in establishing eye-to-eye contact among autistic children and the absence of eye-to-eye contact between mother and baby at birth. It is thanks to my conversations and correspondence with Niko Tinbergen that the concepts of 'primal period' and 'primal health research' came to my mind.

It is also because I met Niko Tinbergen that I read with special attention, in June 1991, a report by Ryoko Hattori, a psychiatrist from Kumamoto, Japan. She had published a thought-provoking study of the risks of autism according to the place of birth, in an authoritative medical journal. She revealed that children born in a hospital where the 'Kitasato University Method' was routine were at increased risk of becoming autistic. Although it does not appear in the title of the article, the main characteristic of this method was labour induction a week before the due date. I considered this Japanese study so important that I found an opportunity to go to Kumamoto in the 1990s and meet Ryoko Hattori.

After this pioneering work, I innocently expected a flood of epidemiological studies on the same topic in the coming years. I expressed my impatience in 2000 by introducing the concept of 'cul-de-sac epidemiology'.[3] Taking the example of autism (and drug addiction), I referred to studies we prefer not to look at, not to enlarge, not to replicate, and not to quote after publication. I'll never know if my paper had a triggering effect, but a long series of valuable studies has been published since 2002, confirming the preliminary conclusions. In 2002, a study was published involving the whole Swedish population born over a period of 20 years (more than 2 million births!). The important point is that, whatever the research protocol or country, the period surrounding birth was always critical. In particular, a 2004 Australian study compared 465 autistic subjects born in Western Australia over a period of 15 years with their 481 siblings and 1,313 controls.

The list of valuable studies of autism from a primal health research perspective is getting longer still. The issue became more complex with the widespread use of the concept of 'autistic spectrum disorders'. A study using health registries of four Nordic countries and Western Australia looked at nearly five million children who had survived their first year of life. In this series there were 671,646 caesarean births and 31,073 children classified as autistic (autistic spectrum disorder). Across the five countries, caesarean births (elective or emergency) appeared as risk factors for autism. The North Carolina study combined data about 625,042 births and school records including the cases of more than 5,500 children diagnosed as autistic. Compared to children born to mothers who received neither labour induction nor augmentation, children born to mothers who were induced and augmented, only induced, or only augmented, were at increased risk for autism after controlling for potential confounders related to socioeconomic status, maternal health, maternal age, pregnancy-related events and conditions, and birth year.

Once more, negative findings must be taken into account to confirm the limits of the critical period. The 2004 Australian study is precious from this viewpoint. Autistic people had the same average birth weight, the same average head circumference at birth and the same placental average weight as the others. Pre-eclampsia did not appear as a risk factor. Two studies of the risks for autistic spectrum disorders among children conceived in vitro (IVF) suggest that the mode of conception does not significantly influence the risks. However, it is possible, according to one of these studies, that the technique of intracytoplasmic sperm injection (ICSI) is associated with a slightly increased risk. This technique, in contrast to conventional IVF, bypasses natural barriers to fertilisation, thereby increasing the possibility of the transmission of genetic defects. Furthermore, what happens after the birth does not significantly influence the risks for autism. The mode of infant feeding does not seem to modify the risks either, and, in spite of highly mediatised theories, we can make the same observation about infant vaccination.

Putting the cart before the horse

This analogy for doing things in the wrong order is appropriate when considering what we know, what we don't know and what is probable about the long-term consequences of early experiences. It is common to conceive theories prematurely, before correlations have been established by epidemiological studies. The issue of autism is typical. All over the world, everybody has heard about theories explaining how infant vaccinations increase the risks of autism. The focus was originally on MMR (the Measles, Mumps, Rubella vaccination) and then it shifted towards vaccines containing a derivative of mercury. Our database cannot present any epidemiological study detecting significant correlations. On the other hand the dozens of valuable studies detecting risk factors for autism at birth have struggled to attract attention in medical circles, let alone from the media and general public. Are theories more attractive than hard data?

While primal health research should be the primary scientific discipline for exploring the long-term consequences of early experiences, it is not self-sufficient. Its limited role is to detect correlations and to evaluate their statistical significance. Its effect is often to stimulate our curiosity in terms of cause and effect. This is why we must evaluate the explanatory power of complementary scientific perspectives. During the last decades of the 20th century, great importance was given to the hormonal system and the concept of 'set point level'. Today the disciplines with the greatest explanatory power are epigenetics and 'metagenomics bacteriology'.

The advent of epigenetics is at the root of a comprehensive rethinking of what makes two individuals different. This emerging discipline is based on the concept of gene expression. Some genes may be allocated a kind of label (an 'epigenetic marker') that makes them silent without altering the DNA sequences. This marker can be a DNA methylation. DNA methylation, which reduces gene expression, is the best-studied epigenetic marker, mainly because tools have existed to study it. The phenomenon of gene expression is influenced by environmental factors, particularly during the primal period. In relation to primal health research, epigenetics gives renewed importance to the concept of critical periods of development, and constitutes a previously missing link between genetic and environmental factors.

We already have a sufficient amount of data at our disposal to claim that, in humans, the perinatal period is a phase of intense epigenetic activity influenced by the mode of birth. A first Swedish study compared DNA methylation in white blood cells after pre-labour caesarean and after birth by the vaginal route.[1] Another study provided similar results by looking at haematopoietic stem cell epigenetics.[2] Neither of these studies considered the case of birth by in-labour caesarean. A British study came to the conclusion that perinatal epigenetic analysis may have utility in identifying individual vulnerability to later obesity and therefore type 2 diabetes.[3] It has been confirmed recently that specific disturbances in DNA methylation predict future development of type 2 diabetes.[4]

Since it appears that epigenetic markers (the

'epigenome') may be to a certain extent transmitted to future generations, we are encouraged to pay renewed attention to studies exploring the transgenerational effects of what happens during the primal period, and include them in our database. Considering these transgenerational effects of early experiences will be another way to enlarge our relationship with time.

We should not ignore the limits of the epigenetic perspective. We are still in a preliminary phase in the exploration of long-lasting differences established during the primal period.

Bacteriology is another fast-developing discipline that can provide plausible interpretations of epidemiological studies. Let us recall that the 'microbiome revolution' is a consequence of technological advances. As long as bacteriologists could only look at microscopes and cultivate microbes on Petri dishes, they could not see the 'unseen majority', since the growth conditions of many microbes are unknown. The turning point started when bacteriologists could dramatically expand their horizons thanks to the power of computer processing and new DNA sequencing technologies. Today human births must be first classified from a bacteriological perspective: there are many intermediate situations between a homebirth and a birth by caesarean in the sterile environment of an operating room, with exposure to antibiotics. Not only is 'metagenomics bacteriology' endowed with an increasing explanatory power, but it should also inspire a new generation of epidemiological studies.

13

A turning point in the size of human groups

The animal kingdom may be classified in a way that transcends the division into mammals, fish, reptiles and birds. Within each class, some species are solitary, while others are gregarious. For example, among feline mammals, tigers are solitary while lions are gregarious. Among apes, members of the chimpanzee family, including Homo, live in groups: they are gregarious. Orangutans, on the other hand, are solitary.

Many students of human nature have tried to evaluate the typical size of original human groups. Robin Dunbar has suggested limits to the number of people with whom one can maintain stable social relationships. By combining several criteria, he proposed that humans can comfortably maintain only 150 stable relationships.[1] Gérard Mendel has raised questions about teamwork, when every participant has a specific and complementary role to play. He often referred to a football (soccer) team, with its 11 players. Interestingly, football is the most popular global sport.[2]

From life in bands to the global village

Before the advent of agriculture and animal husbandry,

our more-or-less nomadic ancestors lived in small groups. Although the social organisation of Paleolithic humans remains largely unknown, we have at our disposal concordant documents suggesting that there were bands ranging from 50 to 150 members (in the region of 'Dunbar's number'). Their horizon was obviously limited, although there is evidence of inter-band trade and exchanges of resources and commodities, such as stone needed for making tools. Furthermore, occasional interbreeding, and therefore 'exchanges of genes' and increased genetic diversification, was an advantage. Valuable sources (recent findings and analyses in evolutionary biology, archaeology, and ethnology) suggest that Palaeolithic societies rarely engaged in organised violence between groups (i.e. war).[3]

Once more, the Neolithic revolution was a spectacular turning point, with the advent of sedentary societies based in built-up villages and towns. From that time, the size of human groups has been constantly increasing. Gradually, there were bigger and bigger villages, towns and cities, until the age of the megalopolis. There were provinces, countries, states and unions of states. However, the deep-rooted human need to belong to a group, whatever its size, never disappeared. The concept of frontier is still valid. Some terms, such as patriotism and respect for the flag, are usually associated with a positive connotation, while others, such as nationalism and chauvinism, are usually associated with a negative connotation.

Homo, a typical gregarious primate, must gradually adapt to the ultimate size in terms of human group. 'Globalisation' was predicted by utopian pioneers. Karl Marx stated that 'working men have no country'. The political slogan 'Proletarians of the world, unite!' is dated 1848. The famous song '*The Internationale*', dated 1864, may be considered a symbol of the turning point with its words 'The Internationale will be the human race'.

Today there is already a de facto unification of humanity. 'Globalisation' has been established in a way that the pioneering idealists of the 19th century could not anticipate. It is related to the adoption of English as a lingua franca, a global network of cheap and fast transportation, the widespread use of the internet, mobile phones and credit cards valid all over the world, new challenges such as global warming and cross-boundary water and air pollution, interdependence of cultural activities and even interdependence of different ways of thinking.

However, in such an apparently unified world, it is not easy to smash all barriers and to think in terms of a global village and the survival of mankind. We need tools to train ourselves to extend our horizons and become familiar with the ultimate collective dimension.

A tool to adapt to the age of globalisation

In the same way that we have presented our database (primalhealthresearch.com) as a tool to train ourselves to think long term, we can also present it as a tool to train ourselves to adapt to the age of globalisation. An enlarged collective dimension is a characteristic of the 21st-century lifestyle in general and medicine in particular. When I was a medical student, in the middle of the 20th century, most teachers were still authoritative and respected practitioners who were constantly referring to personal experience, anecdotes and fashionable theories. '*Eminence*-based medicine' has been taught for thousands of years. This is how medical practices, such a bloodletting, clysters, cupping glasses, or 'lying-in' during the two or three weeks following birth have been transmitted from generation to generation.

Although John Snow may be considered the father of epidemiology after his work tracing the source of the 1854 cholera epidemic in London, it was only during the last decades of the 20th century that 'evidence-based medicine' developed, thanks to the use of computer processing and spectacular advances in statistical analysis. It became easier to publish sophisticated studies based on huge numbers.

The primal health research database is an ideal source of documents for becoming familiar with the collective dimension. Some keywords are particularly fruitful in highlighting studies based on enormous numbers. This is the case, for example, of the keyword 'autism'. A study published in 2002 involved all Swedish children born during a period of 20 years (from 1974 to 1993) and all children discharged from a Swedish hospital diagnosed with autism.[4] A multi-national study of caesarean section and the risk of autism involved 5 million births.[5] An Australian study featured subjects born in Western Australia between 1980 and 1995 and diagnosed with an autistic spectrum disorder by 1999.[6] A study of autistic spectrum disorders among children born in Kaiser Permanente hospitals in South California between 1991 and 2009 analysed nearly 600,000 medical records.[7] The North Carolina study of the association of autism with induced or augmented labour included more than 5,500 autistic children.[8]

Many other keywords can help us to become familiar with huge numbers. For example 'forceps' opens the way to a study of 52,000 draftees born in Jerusalem between 1964 and 1972. At the age of 17, they had intelligence tests before entering the army.[9] A search for 'vaccination' reveals that, while the media were focusing on anecdotes suggesting that measles, mumps, and rubella vaccination in infancy might be a risk factor for autism, epidemiologists were analysing 537,000 children born in Denmark between 1991 and 1998.[10] Even for restricted topics such as the long-term effects of antibiotics prescribed for preterm rupture of the membranes, or for spontaneous preterm labour, researchers looked at groups of more than 4,000 children.[11,12]

The cultural mammal

Apart from the gregarious nature of humans, there is another reason to give special importance to the collective dimension where our species is concerned. Because human beings speak, develop sophisticated ways to communicate and create cultural milieus, there are situations where human behaviours are less directly under the effects of the hormonal balance and more directly under the effects of the cultural milieu. This is the case in pregnancy and childbirth. When a woman is pregnant, because she is a human being she can express through language that she is expecting a baby and anticipate maternal behaviour. Other mammals cannot do that. They have to wait until the day when they release a cocktail of love hormones to be interested in their babies. In non-human mammals, when the birth process is disturbed, the effects are spectacular and easily detected at an individual level: as a general rule, the mother is not interested in her baby. This is the case, for example, of ewes giving birth with epidural anaesthesia[13] or monkeys giving birth by caesarean: the babies can survive only if human beings take care of them.[14] Millions of women, on the other hand, take care of their newborn babies in spite of powerful interferences.

We should not conclude that we have nothing to learn from other mammals. They suggest which questions we should raise where human beings are concerned: with these questions we must always introduce the collective dimension via words such as 'civilisation' and even 'humanity'. This is why the current main questions are about the future of a humanity born by caesarean, or with epidural anaesthesia, or with drips of synthetic oxytocin.

14

A turning point in the history of natural selection

The current crisis

The mystery surrounding breech presentation at term has been suddenly clarified. This is one reason among others to raise questions about the probable evolutionary effects of modern obstetrics.

Countless theories have been proposed over the years to explain why, in all ethnic groups, a small number of babies are born at term buttocks first or feet first and why, whatever the morphological particularities of a population, the rate of breech births is mysteriously similar, with an upper limit of around 3%. It has always been considered a serious issue, because everywhere and in every phase of human history, a breech presentation was deemed dangerous and inauspicious.[1,2] In many cultures, methods to try to turn babies so that they are born head first have been transmitted from generation to generation. This is a confirmation of the universal bad reputation of breech birth. In traditional Chinese medicine, they tried to turn babies by associating 'moxibustion' (burning dried mugwort on an acupuncture point on the last toe) and chest-knees postures. In Europe, 'external version' was practised in antiquity. Via films and written documents, Brigitte Jordan provided details of the use of this manoeuvre by traditional Yucatan midwives.[3] We even have documents suggesting that breech birth is also comparatively dangerous in pre-Neolithic societies where women isolate themselves to give birth: in his book about life among the Pirahã, in the Amazonian jungle, the only case of maternal death mentioned by Daniel Everett related to a breech birth.[4]

The publication of a huge Norwegian study has shed more light on the issue. The authors used the national birth registry, based on all births in Norway from 1967 to 2004 (2.2 million births).[5] Data were provided by about 450,000 mother-baby pairs and about 300,000 father-baby pairs. An original and fruitful aspect of the study was the identification of 'paternal half-siblings' (siblings with the same father and different mother). They identified more than 35,000 paternal half-siblings whose father had changed partner between the births of his first two children, and where both siblings were the first born of the two mothers. This is how the authors could conclude that having a breech-born father or a breech-born mother more than doubled the child's risk of being breech.

Plausible mechanisms suggest that the main factors influencing the presentation of the baby at birth are genetic, with maternal or paternal transmission. If this is confirmed, it will be an important piece of news in the context of the 21st century, at a time when the dominant strategy, all over the world, is routine caesarean section in the case of a breech presentation at term. The effect of this strategy is that, from now on, a breech birth is as safe as a head-first birth for the mother and for the baby. For example, in a Canadian series of 46,766 planned caesareans for breech presentation at term (not for maternal pathological conditions) no mother died.[6] In the well-known *Lancet* study which has been instrumental in establishing the new doctrines, the 'perinatal mortality rate' was 1.6%: even for a population of babies born head first, this rate would be considered low.[7]

With the advent of easy, fast and safe techniques of caesarean, the genetic predisposition to breech presentation is more easily transmitted to the following generations. Theoretically, in the future the rate of breech birth will not be maintained at around 3% and instead will gradually increase. We might even suggest, although many associated factors should be considered, that the tendency towards more breech births at term is already detectable. The proportion of breech births in the Norwegian registry was 2.5% in 1967–76, 3.0% in 1977–86, 3.2% in 1987–96, and 3.5% in 1997–2004... With the data we already have at our disposal, we can imagine that, with the help of appropriate methods of computer programming, it should be possible to establish projections to evaluate the prevalence of breech births one or two centuries from now. Today being born breech is a way to belong to the increasing population of human beings who are dependent on medicine (and health budgets) from the beginning of their life.

We took the example of breech presentation at term to illustrate the concept of neutralised laws of natural selection. We might have chosen another point of departure to show that, from now on, if a woman cannot give birth easily and safely by herself, she may nonetheless have the same number of children as other mothers. When taking into account the great diversity of potentially dangerous births, we realise the magnitude of the topic. Can a better understanding of physiological processes slow down an apparently irreversible process?

We must also raise questions about pharmacological assistance, particularly the use of synthetic oxytocin, as a possible way to modify the current tendencies. Until now, the possible long-term effects of the different components of pharmacological assistance have not been seriously studied. They are probably underestimated. Anyway, at the current turning point in the history of childbirth, we have already reached a time when the number of women who give birth to babies and placentas thanks only to the release of natural hormones is becoming insignificant. In other words, love hormones have been made redundant in the perinatal period.

The process of selection is also modified by reproductive medicine at the phase of conception. Today very fertile people don't have significantly more children than those who are sterile or sub-fertile. Medicalised contraception – which became available in the 1970s – is a recent and easy way to control the number of children per person. On the other hand, many cases of female or male sub-fertility or infertility are treated by medically assisted conception. In other words, the number of children born per person depends on other factors than the degree of fecundity. This is another essential aspect of the history of our species.

The previous crisis

Once more, to understand the current crisis we need to refer to the Neolithic revolution, the previous turning point characterised by the advent of agriculture and animal husbandry. It is through the 'artificial selection' of plants and animals that our ancestors started to dominate nature. It is worth recalling that Darwin was interested in the process of 'artificial selection' as an illustration of the concept of natural selection. Key species of plants such as wheat and rice in Eurasia, and maize in Mexico, became significantly different from their wild ancestors. In the same way animal husbandry relied on the transmission of traits considered beneficial, such as growth rate, survival rate, meat quality, age at sexual maturation and fecundity.

It is in such a context that we must consider the particular case of Homo. From what we know about ethnic groups which, until recently, had kept Palaeolithic characteristics, we can conclude that a cultural regulation of genital sexuality is an aspect of the domination of nature. While the number of children a woman can have between puberty and menopause has always been limited for obvious physiological reasons, we can assume that, before the organisation of mating, some dominant and very fertile men were transmitting far more genes than others. The laws of natural selection were not significantly altered in pre-Neolithic human groups. Sedentary lifestyles and matrimonial arrangements based on cultural criteria and rules changed that situation.

Tougher selection

The socialisation of childbirth, as an aspect of the domination of nature, had complex effects on the process of selection. The main effect was to make it tougher. A cultural conditioning was gradually established, suggesting a woman needs help to give birth. While originally women were *giving birth*, a phase of human history was reached when babies were supposed *to be delivered* by a birth attendant. Whatever the cultural variants, the need for privacy was denied. With increased socialisation, human births became more and more difficult, and therefore more and more dangerous for mothers and babies. This was a way to make selection tougher.

The transmission of perinatal beliefs and rituals from generation to generation is another aspect of the socialisation of childbirth. It is another way to reinforce the laws of natural selection. The immediate effect of some rituals was obviously to interfere with the process of selection by eliminating those who were considered too weak to become members of the community. This is the case, for example, of the widespread ritual of immersing babies in icy water: the weak ones could not survive.[2]

In general, the elimination of weak babies was not a conscious objective: it was a subtle effect of perinatal beliefs and rituals. For thousands of years, in all human societies we know about, mothers and newborn babies have been separated and the initiation of breastfeeding has been delayed. In other words, it has long been routine to neutralise the 'maternal protective aggressive instinct'. The nature of this universal mammalian instinct is easily understood when one imagines, for example, what would happen if one tried to pick up the newborn baby of a mother gorilla who has just given birth.

It would take volumes to review all the invasive perinatal beliefs and rituals that have been reported. As early as 1884, *Labor Among Primitive Peoples* by George Engelmann provided an impressive catalogue of the countless ways of interfering with the first contact between mother and newborn baby. It described beliefs and rituals occurring in hundreds of ethnic groups on all five continents.[8]

The most universal and intriguing example of cultural interference is simply to promote the belief that colostrum is tainted or harmful to the baby, and that it needs to be expressed and discarded.[9] The negative attitude towards colostrum implies that, immediately after the birth, a baby must be in the arms of another person, rather than with his or her own mother. This is related to the widespread deep-rooted ritual of cord-cutting soon after birth.[10] Several beliefs and rituals are often associated, all of them reinforcing each other. In Western Europe, these universal rules could be explored through historical, rather than anthropological, documents.[11] When referring to the history of orgasmic states, we have already underlined the similarities between two European regions: Brittany, and Tudor and Stuart England.

Turning point or U-turn?

We are suddenly experiencing more than a turning point in the history of the socialisation of childbirth: it is a U-turn that started during the second half of the 20th century with the advent of modern obstetrics. After thousands of years of reinforced selection, we are entering the phase of neutralised selection. The number of children per woman depends on other factors than the capacity to give birth. Should we start raising questions about the possible effects of the current U-turn on the evolution and the survival of our species?

These issues will become still more complex in the near future, when we'll dare to consider the possible evolutionary effects of emerging branches of reproductive medicine: medicalised contraception, medically assisted conception and pre-natal diagnosis in particular.

15

A turning point at two and a half minutes before midnight

The famous English theoretical physicist Stephen Hawking once posed an open question: 'In a world that is in chaos politically, socially and environmentally, how can the human race sustain another 100 years?' When asked to clarify his point of view, he replied: 'I don't know the answer. That is why I asked the question, to get people to think about it, and to be aware of the dangers we now face'.

As early as 1947, the members of 'The Bulletin of the Atomic Scientists' introduced the concept of the 'Doomsday clock', a symbolic countdown to 'midnight' – the end of humanity. These scientists had prestigious sponsors and advisers including Nobel laureates. The clock's original setting was seven minutes to midnight. It has been set backward and forward many times since then. In January 2017, it was set at two and a half minutes to midnight.[1] Originally, a global nuclear war was thought to be the main threat to the survival of mankind. Recently, the effects of other possible threats related to human activities have been added to the list. Today, a team led by Dennis Pamlinof of the Global Challenge Foundation and Stuart Armstrong of the Future of Humanity Institute is offering an enlarged list of global risks with a potentially infinite impact.

Until now, scientists thinking about the future have mostly been physicists. They have considered threats related to human activities, but they have not taken into account probable transformations of Homo under the effects of radically new lifestyles. In the near future there will probably be several spectacular transformations of our species. Some of these transformations might be the effects of an increased genetic diversity as a consequence of fast and cheap transcontinental flights. Others might be related to frequent and prolonged exposure to artificial light, an essential aspect of 21st-century lifestyle: melatonin (the darkness hormone) has strong links with the other neuromodulators, including oxytocin.

A suggestion

As a first step, in a dialogue with the scientists in charge of these predictive bulletins, we suggest that the neutralisation of the laws of natural selection by obstetrics and other branches of reproductive medicine should be given a prominent place as an apparently irreversible factor: it tends to set forward the 'Doomsday clock'. It is urgent to realise the probable and unprecedented evolutionary effects of easy, fast and safe techniques of caesarean section.

Since physicists are familiar with the language of diagrams, I have designed for them a simplified line graph related to the birth of Homo. It is about the history of the process of reinforcement-neutralisation of the laws of evolution through selection. The objective is to illustrate the spectacular U-turn that started during the second half of the 20th century. The point 0 refers to the emergence of the genus Homo. The point 1 refers to the Neolithic revolution. The point 2 indicates the U-turn. After the Neolithic revolution, some aspects of the laws of selection were first reinforced (+) and then neutralised (–).

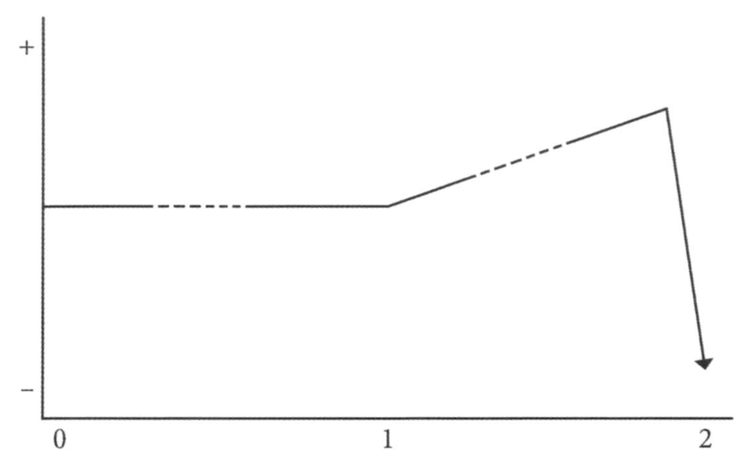

A recent spectacular U-turn.

Although this diagram was designed for scientists in charge of the Doomsday clock, it might attract the attention of groups and individuals who are raising questions about the future. Consider Avaaz, arguably the globe's largest and most powerful activist network. According to its director and co-founder Ricken Patel, its mission is to 'close the gap between the world we have and the world most people everywhere want': 'Idealists of the world, unite!' is a parody of 'Proletarians of the world, unite!', the political slogan from 1848. Consider too Mark Zuckerberg, the founder of Facebook, who wrote in his 2017 manifesto, that 'Progress now requires humanity coming together not just as cities or nations, but also as a global community'.

The scientists, the idealists and the fictionists who are endowed with the capacity to look towards the future share a common failing: they do not take into account probable and spectacular transformations of Homo in relation, in particular, to the emergence of reproductive medicine. Is it premature to suggest that the future of the planet might be first and foremost dependent on the evolution of Homo?[2]

Our diagram should also attract the attention of evolutionists. Some of them have introduced the concept of 'evolutionary thinking in medicine'.[3] It is notable that the common point between most evolutionists is still a lack of interest in the future of Homo as a species: the 'scars of evolution' have been their principal topic.[4]

Combining evolutionary thinking and interest in the future

Today our preliminary objective should be to develop in parallel evolutionary thinking and interest in the future. This implies a deep process of reconditioning. This would lead to questions inspired by emerging scientific disciplines. From the middle of the 20th century, evolutionary thinking was dominated by Neo-Darwinism: the focus was on the transmission of genes and, occasionally, of genetic mutations. In the age of epigenetics, at a time when there is a renewed interest in theories that preceded Darwin, some facts have become acceptable since they are open to interpretation. In particular it is becoming acceptable to claim that when organs or physiological functions are underused they become weaker from generation to generation.

There is one spectacular example of a human physiological system that has become suddenly less useful. It is the oxytocin system. During the birth process, its activation is supposed to reach another order of magnitude than during any other situation, even sexual intercourse and breastfeeding. At a time when synthetic oxytocin is cheap and widely used across the globe, and at a time when the caesarean section has become an easy, fast and safe operation, the number of women who rely on their oxytocin system to give birth to babies and placentas is becoming insignificant. We must also keep in mind that a milk ejection reflex is induced by oxytocin release. After taking into account the current average number of babies per woman and the very short average duration of breastfeeding, it is easy to reckon that the number of milk ejection reflexes in the life of a modern woman is very small compared with what it has been in other societies: the oxytocin system is also underused to feed babies.

Should we expect that the probable transformations of Homo might include a weakening of the oxytocin system? I dare to go a step further and suggest that perhaps this physiological system is already deteriorating. We can express concern about this possibility by bringing together a great diversity of published data. With regards to childbirth, consider an American study entitled 'Changes in labor patterns over 50 years'.[5] The authors compared a first group of nearly 40,000 births that occurred between 1959 and 1966 and a second group of nearly 100,000 births that occurred between 2002 and 2008. They only looked at births of one baby at term, with head presentation and spontaneous initiation of labour. After taking into account many factors such as the age, height and weight of the mother, it appeared that the duration of the first stage of labour was dramatically longer in the second group. It was two hours and a half hours longer in the case of a first baby, and two hours in the other cases. For practitioners of my generation this study demonstrates the obvious: even before being aware of such data, I was personally convinced that the human capacity to give birth is already deteriorating.

Countless other studies suggest that in spite of intense public health campaigns to promote breastfeeding, and although contemporary populations are better informed than ever about the irreplaceable value of mother's milk, recent breastfeeding statistics are worrying. As for genital sexuality, many factors are involved and hard data are not available, but most sex therapists are overworked and drugs to correct sexual dysfunction are at the top of the lists of pharmaceutical substances in terms of commercial value.

The future of the capacity to love

It is well accepted that the oxytocin system is involved in all facets of love: so what happens to love when the oxytocin system goes unused? The capacity for empathy, as a facet of love, is a personality trait that has been evaluated through scores developed by modern psychological sciences. At the annual meeting of the Association for Psychological Science in June 2010, a synthesis of 72 studies of the evolution of personality traits of American students ('College graduates') between 1979 and 2009 was presented.[6] According to this research, college graduates are 40% less empathetic than those of decades past. The decline has been progressive, particularly after the year 2000. Once more, this is compatible with a declining oxytocin system.

While it appears unrealistic – even impossible – to re-establish the laws of natural selection in humans, we may wonder whether there are ways to slow down the clock before we reach midnight. What can we do so that oxytocin – as the main component of all cocktails of 'love hormones' – can be intensely released as early as possible during human life? Would it be possible, for example, to overcome our dominant conditioning? Would it be possible to spread the word and convince anyone that, in terms of reproductive physiology, human beings are mature before the age of 20? In other words, can we make it 'fashionable' to have a baby at age 18? What would be the effects of combining such a new 'fashion' with a renewed understanding of birth physiology?

These considerations about probable transformations of Homo led us to dramatically deviate from the initial preoccupations of the atomic scientists in the aftermath of the Second World War. Today we can simply observe that humanity is still alive and that the scientists of the 1940s had underestimated the complexity of their ambitious project.

Conclusion

Clarification: Who is Homo?

An enlarged family

Three centuries ago, when Carl Linnaeus formalised the modern way of naming living organisms, he did not bother to define Homo. He had no reason to think that the genus Homo would ever have additional members. The only species belonging to the genus Homo was called 'sapiens'. This is why, even today, the term Homo has not been properly defined.[1] It may be used in an enlarged sense, referring to all the hominids that separated from the other members of the chimpanzee family (the genus Pan) about six million years ago. It may be used with a restricted sense, as a synonym for modern human beings.

Throughout this book, when using the word Homo, we have been referring to hominids characterised by a huge brain, more precisely those with an adult brain volume far above 1,000 cubic centimetres... another order of magnitude compared with all others members of the chimpanzee family. This includes modern humans and other variants of the known or not-yet-known 'big-brained hominids' such as Neanderthals and Denisovans. Today it is becoming difficult to separate these variants of the 'big-brained chimpanzee', since we are accumulating data concerning the multiple and diverse lasting effects of interbreeding. We could already claim that Neanderthals had not disappeared, since modern humans of Asian and European origin share Neanderthal gene variants that are associated with physiological particularities. But what about Denisovans?

Kept in a freezer

Focusing on Denisovans leads us to realise how evolutionary sciences have been revolutionised by genetics. The intense curiosity about Denisovans was aroused by genetic perspectives. The only fossils scientists have at their disposal so far are a finger bone fragment from a juvenile female, two teeth, and a toe bone. The oldest publication related to these fossils is dated 2010.[2] The fossils were found in the remote Denisova cave in the Altai Mountains in South Siberia, a cave that has also been inhabited by Neanderthals and Sapiens. This place is a paradise for evolutionary geneticists, because the average temperature inside the cave is 0°C. It is as if the precious fossils had been kept in a freezer.

From the studies already published, we can conclude that there has been interbreeding of Denisovans with Neanderthals, Sapiens, and probably another as-yet-unidentified Homo. It is also well established that today about 3% to 5% of the DNA of Melanesians and Aboriginal Australians derives from Denisovans.[3,4] The observation of high Denisovan admixture in Oceania rather than in mainland Asia is supportive of the concept of the 'marine chimpanzee'. After the preliminary published studies, new data suggest that Denisovan ancestry is more widespread in modern populations than previously thought.[5] We are now expecting revealing questions to begin to be answered, such as questions about the origin of the documented Denisovan contribution in the genome of American Indians: we have mentioned the reasons to consider plausible the use of the South Pacific route by the 'marine chimpanzee'. There have even been studies of the epigenome of Denisovans, but epigenetic alterations can vary drastically between tissues, so epigenomes gathered from bones and teeth will not say anything about organs such as the brain.[5]

Since 2014 a new generation of research has led to the conclusion that the genetic legacy of Denisovans is associated with observable characteristics (the 'phenotype') of some modern humans. In this new research framework we must first mention a study of the adaptation of Tibetans to the hypoxic environment of high altitude plateaus: a hypoxia pathway gene is found only in Denisovans and Tibetans.[7] This can be convincingly explained by the transfer of genetic information from Denisovans to Sapiens. It is probable that, partly thanks to gene variants provided by Denisovans, Greenland Inuit are adapted to the cold through their heat-generating body fat.[8] It also appears that Denisovan genes are linked to a more subtle sense of smell, particularly in Papua New Guineans.[9] Furthermore, we must attach great importance to preliminary studies suggesting that the transfer of genetic information from Denisovans (and Neanderthals) can explain adaptive variations among modern humans to pathogens such as bacteria, fungi, and parasites.[10]

A turning point in cynical humour

At a time when emerging and fast-developing scientific disciplines offer renewed perspectives to explore the complexity of human nature, I suggest that, in the particular case of our species, we abandon the conventional binomial classification. We only need the unifying term 'Homo' when referring to the big-brained members of the chimpanzee family. How frustrating for the cynical humorists who make fun of the adjective 'sapiens'!

References

Chapter 2: A turning point in our understanding of the human condition

1. Venturi, S., Venturi, M. 'Iodine, PUFAs and Iodolipids in Health and Diseases: Evolutionary perspective'. In: *Human Evolution*. Angelo Pontecorboli Editore Firenze. 2014:29: 185-205
2. Schone, F., Sporl, K., Leiterer, M. 'Iodine in the feed of cows and in the milk with a view to the consumer's iodine supply'. *J Trace Elem Med Biol.* 2017 Jan;39:202-209. doi: 10.1016/j.jtemb.2016.10.004. Epub 2016 Oct 11.
3. Monahan, M., Boelaert, K., Jolly, K., et al. 'Costs and benefits of iodine supplementation for pregnant women in a mildly to moderately iodine-deficient population: a modelling analysis'. *Lancet Diabetes Endocrinol* 2015 Aug 7. pii: S2213-8587(15)00212-0. doi: 10.1016/S2213-8587(15)00212-0. [Epub ahead of print]
4. Guess, K., Malek, L., Anderson, A., et al. 'Knowledge and practices regarding iodine supplementation: A national survey of healthcare providers'. *Women Birth* 2016 Sep 2. pii: S1871-5192(16)30094-4. doi: 10.1016/j.wombi.2016.08.007. [Epub ahead of print]
5. Zygmund, A., Lewinski, A. 'Iodine prophylaxis in pregnant women in Poland – where we are?' (update 2015). *Thyroid Research* 2015 Dec 8;8:17. doi: 10.1186/s13044-015-0029-z. eCollection 2015.
6. Weyrich, L.S., Duchene, S., et al. 'Neanderthal behaviour, diet, and disease inferred from ancient DNA in dental calculus'. *Nature* 2017 Mar 8. doi: 10.1038/nature21674. [Epub ahead of print]
7. Mone, F., Mulcahy, C., McParland, P., McAuliffe, F.M. 'Should we recommend universal aspirin for all pregnant women?' *Am J Obstet Gynecol* 2016 Sep 19. pii: S0002-9378(16)30804-3. doi: 10.1016/j.ajog.2016.09.086. [Epub ahead of print]
8. Vane, John Robert 'Inhibition of prostaglandin synthesis as a mechanism of action for aspirin-like drugs'. *Nature: New Biology* 1971; 231 (25): 232–5
9. Ikeda, Y., Shimada, K., Teramoto, T., et al. 'Low-dose aspirin for primary prevention of cardiovascular events in Japanese patients 60 years or older with atherosclerotic risk factors: a randomized clinical trial'. *JAMA* 2014 Dec 17;312(23):2510-20. doi: 10.1001/jama.2014.15690.
10. http://www.vin.com/apputil/content/defaultadv1.aspx?pId=11331&meta=Generic&id=4848030&print=1 1/1
11. Odent, Michel 'Obstetrical implications of the aquatic ape hypothesis' In: *Was Man More Aquatic in the Past?* Marc Verhaegen, Algis Kuliukas, Mario Vaneeechoutte ed. Chapter 9. Bentham Science, 2009.
12. Jenkins, D.T., Wysocki, S.J., Davies, D.M. 'Amniotic fluid squalene: a useful test in prolonged pregnancy'. *Aust NZ J Obstet Gynaecol* 1982 Aug;22(3):135-7
13. Williams, Marcel Francis 'Marine adaptations in human kidneys' In: *Was Man More Aquatic in the Past?* Marc Verhaegen, Algis Kuliukas, Mario Vaneeechoutte ed. Chapter 8. Bentham Science, 2009.
14. Hardy, Alister *Was Man More Aquatic in the Past? New Scientist* 1960; 7: 642-5.
15. Morgan, Elaine *The Descent of Woman*. Souvenir Press, London, 1972.
16. Morgan, Elaine. *The Aquatic Ape*. Souvenir Press, London, 1982.
17. Morgan, Elaine. *The Scars of Evolution*. Souvenir Press, London, 1990.
18. Crawford, M.A., Marsh, D. *The Driving Force*. William Heinemann, London, 1989.
19. Cunnane, Stephen. *Survival of the Fattest*. World Scientific Publishing, Singapore, 2005
20. Alberts, S.C., Altmann, J., et al. 'Reproductive aging patterns in primates reveal that

humans are distinct'. *Proc Natl Acad Sci USA*. 2013 Aug 13;110(33):13440-5. doi: 10.1073/pnas.1311857110. Epub 2013 Jul 29.

21. Croft, D.P., Brent, L.J., et al. 'The evolution of prolonged life after reproduction'. *Trends Ecol Evol*. 2015 Jul;30(7):407-16. doi: 10.1016/j.tree.2015.04.011. Epub 2015 May 14.
22. Brent, L.J.N., Franks, D.W., Foster, E.A., Balcomb, K.E., Cant, M.A., and Croft, D.P. 'Ecological knowledge, leadership, and the evolution of menopause in killer whales'. *Curr. Biol.* 2015 25, 746–750. doi: 10.1016/j.cub.2015.01.037
23. Johnson, Jessica, Odent, Michel. *We are all Water Babies*. Dragon's World, London, 1994.
24. Pedersen, M.W., Ruter, A., Schweger, C., et al. 'Postglacial viability and colonization in North America's ice-free corridor'. *Nature* September 2016;537: 45-49.
25. Collins, M., Dillehay, T. 'Early cultural evidence from Monte Verde in Chile'. *Nature* 1988; 332: 150-152.
26. Skoglund, P., Mallick, S., et al. 'Genetic evidence for two founding populations of the Americas'. *Nature* 2015 July 21 .doi: 10.1038/nature14895
27. *Nature*, DOI: 10.1038/nature 08844
28. Yirka, Bob. 'Evidence suggests Neanderthals took to boats before modern humans'. 1 March 2012, phys.org. Retrieved 5 May 2016.
29. Marshall, Michael. 'Neanderthals were very ancient mariners' *New Scientist* 29 February 2012. Retrieved 5 May 2016.
30. Charles, Q. 'Ancient Mariners: did Neanderthals sail to Mediterranean islands?' *LiveScience* 15 November 2012. Retrieved 5 May 2016.
31. *Journal of Archaeological Science*, DOI: 10.1016/j.jas.2012.01.032.
32. Hornell, James. 'The Role of Birds in Early Navigation'. *Antiquity* September 1946; 20 (70): 142-149. DOI:http://dx.doi.org/10.1017/S0003598X0001953031-
33. Holen, S.R., Deméré, T.A., Fisher, D.C., et al. 'A 130,000-year-old archaeological site in southern California, USA'. *Nature* April 2017; 544: 479-483. doi:10.1038/nature22065
34. Sion, Hopfe, C., Weib, C.L. 'Neandertal and Denisovan DNA from Pleistocene sediments'. *Science* 27 Apr 2017: DOI: 10.1126/science.aam9695

Chapter 3: A turning point in our understanding of human birth

1. Roberts, E. and Frankel, S. 'γ-Aminobutyric acid in the brain'. *Federation Proceedings* 9:219, 1950.
2. Roberts, E. 'Disinhibition as an organizing principle in the nervous system: the role of the GABA system'. In: Roberts E., Chase, T.N., Tower, D.B., Eds. *GABA in nervous system function*. New York: Raven, 1976:515–539.
3. Endevelt-Shapira, Y., Shushan, S., Roth, Y., Sobel, N. 'Disinhibition of olfaction: human olfactory performance improves following low levels of alcohol'. *Behav Brain Res*. 2014 Oct 1;272:66-74. doi: 10.1016/j.bbr.2014.06.024. Epub 2014 Jun 25.
4. Odent, M. 'The early expression of the rooting reflex'. *Proceedings of the 5th International Congress of Psychosomatic Obstetrics and Gynaecology*, Rome 1977. London: Academic Press, 1977: 1117-19.
5. Odent M. 'L'expression précoce du réflexe de fouissement'. In: *Les cahiers du nouveau-né*. Paris 1978; 1-2: 169-185
6. Marlier, L., Schaal, B., Soussignan, R. 'Orientation responses to biological odours in the human newborn. Initial pattern and postnatal plasticity'. *C R Acad Sci III*. 1997 Dec;320(12):999-1005. PMID: 9587477 [PubMed - indexed for MEDLINE]
7. Varendi, H., Porter, R.H., Winberg, J. 'The effect of labor on olfactory exposure learning within the first postnatal hour'. *Behav Neurosci*. 2002 Apr;116(2):206-11.

PMID: 11996306 [PubMed - indexed for MEDLINE]

8. Cernoch, J.M., Porter, R.H. 'Recognition of maternal axillary odors by infants'. *Child Dev.* 1985 Dec;56(6):1593-8.
9. Sullivan, R.M., Toubas, P. 'Clinical usefulness of maternal odor in newborns: soothing and feeding preparatory responses'. *Biol Neonate.* 1998 Dec;74(6):402-8.
10. Varendi, H., Christensson, K., Porter, R.H., Winberg, J. 'Soothing effect of amniotic fluid smell in newborn infants'. *Early Hum Dev.* 1998 Apr 17;51(1):47-55
11. Varendi, H., Porter, R.H., Winberg, J. 'Attractiveness of amniotic fluid odor: evidence of prenatal olfactory learning?' *Acta Paediatr.* 1996 Oct;85(10):1223-7
12. Schaal, B., Marlier, L., Soussignan, R. 'Olfactory function in the human fetus: evidence from selective neonatal responsiveness to the odor of amniotic fluid'. *Behav Neurosci.* 1998 Dec;112(6):1438-49
13. Marlier, L., Schaal, B., Soussignan, R. 'Neonatal responsiveness to the odor of amniotic and lacteal fluids: a test of perinatal chemosensory continuity'. *Child Dev.* 1998 Jun;69(3):611-23
14. Marlier, L., Schaal, B., Soussignan, R. 'Bottle-fed neonates prefer an odor experienced in utero to an odor experienced postnatally in the feeding context'. *Dev Psychobiol.* 1998 Sep;33(2):133-45
15. McGraw, M.B. 'Swimming Behavior of the Human Infant'. *Journal of Pediatrics* 1939;15:485-90.
16. Johnson, J., Odent, M. 'What newborn babies can do'. In: *We Are All Water Babies.* Dragon's World, 1994.
17. Orefice, G., Modafferi, N., et al. 'Archaic reflexes in normal elderly people'. *Acta Neurol* 1991 Feb;13(1):19-24
18. Wang, F., Li, J., et al. 'The GABA(A) receptor mediates the hypnotic activity of melatonin in rats'. *Pharmacol Biochem Behav* 2003 Feb;74(3):573-8.
19. Tysio, R., Nsardou, R., et al. 'Oxytocin-mediated GABA inhibition during delivery attenuates autism pathogenesis in rodent offspring'. *Science.* 2014 Feb 7;343(6171):675-9. doi: 10.1126/science.1247190.
20. Cohen, M., Roselle, D., Chabner, B., Schmidt, T.J., Lippman, M. 'Evidence for a cytoplasmic melatonin receptor'. *Nature.* 1978; 274:894-895.
21. Sharkey, James Thomas, 'Melatonin Regulation of the Oxytocin System in the Pregnant Human Uterus' (2009). *Electronic Theses, Treatises and Dissertations.* Paper 1791. http://diginole.lib.fsu.edu/etd/1791
22. Olcese, J., Beesley, S. 'Clinical significance of melatonin receptors in the human myometrium'. *Fertil Steril* 2014 Jul 8. pii: S0015-0282(14)00566-4. doi:10.1016/j.fertnstert.2014.06.020. [Epub ahead of print]
23. Schlabritz-Loutsevitch, N., Hellner, N., Middendorf, R., Muller, D., Olcese, J. 'The human myometrium as a target for melatonin'. *J Clin Endocrinol Metab.* 2003;88(2):908-913.
24. Sharkey, J.T., Puttaramu, R., Word, R.A., Olcese, J. 'Melatonin synergizes with oxytocin to enhance contractility of human myometrial smooth muscle cells'. *J Clin Endocrinol Metab* 2009 Feb;94(2):421-7. doi: 10.1210/jc.2008-1723. Epub 2008 Nov 11.
25. Bagci, S., Berner, A.L., et al. 'Melatonin concentration in umbilical cord blood depends on mode of delivery'. *Early Human Development* 2012; 88(6):369-373
26. dumas.ccsd.cnrs.fr/dumas-01412390
27. Ingeborg Stadelmann. *Die Hebammen Ssprechstunde.* Stadelmann Verlag, 2006
28. Sharp, K., Brindle, P.M., et al. 'Memory loss during pregnancy'. *Br J Obstet Gynaecol* 1993;100(3):209-15.
29. Brindle, P.M., Brown, M.W., et al. 'Objective and subjective memory impairment in

pregnancy'. *Psychol Med* 1991 Aug;21(3):647-53.
30. Odent, M. 'Fear of death during labour'. *Journal of Reproductive and Infant Psychology* 2001;9:43-47
31. Newton, N., Foshee, D., Newton, M. 'Experimental inhibition of labor through environmental disturbance'. *Obstet Gynecol* 1966;67:371-377.
32. Newton, N. 'The fetus ejection reflex revisited'. *Birth* 1987;14(2):106-108.
33. Odent, M. 'The fetus ejection reflex'. *Birth* 1987;14(2):104-105.
34. Odent, M. 'New reasons and new ways to study birth physiology'. *Int J Gynecol Obstet.* 2001;75 Suppl 1: S39-S455
35. Nissen, E., Lilja, G., Widström, A.M., Uvnäs Moberg, K. 'Elevation of oxytocin levels early post partum in women'. *Acta Obstet Gynecol Scand.* 1995 Aug;74(7):530-3

Chapter 4: A turning point in the history of birth preparation

1. Winnicott, D.W., 1956. 'Primary Maternal Preoccupation' (10) to (20) are included in: *Collected Papers: Through Paediatrics to Psycho-Analysis* (London: Tavistock 1958)
2. Rendell, P.G., Henry, J.D. 'Prospective-memory functioning is affected during pregnancy and postpartum'. *J Clin Exp Neuropsychology* 2008 Nov;30(8):913-9. doi: 10.1080/13803390701874379. Epub 2008 Mar 14
3. Hoekzema, E., Barba-Muller, E., et al. 'Pregnancy leads to long-lasting changes in human brain structure'. *Nat Neurosci.* 2016 Dec 19. doi: 10.1038/nn.4458. [Epub ahead of print]

Chapter 5: A turning point in the evolution of brain size

1. Dominoni, D.M., Goymann, W., Helm, B., Partecke, J. 'Urban-like night illumination reduces melatonin release in European Blackbirds (*Turdus merula*)' *Front Zool.* 2013; 10: 60. doi:10.1186/1742-9994-10-60.
2. Cooke, R.W., Lucas, A., Yudkin, P.L., Pryse-Davies, J. 'Head circumference as an index of brain weight in the fetus and newborn'. *Early Hum Dev.* 1977;1:145–149.
3. Kruska, D. 'Mammalian domestication and its effect on the brain structure and behavior'. In: *Intelligence and Evolutionary Biology*, 211-250. Jerison, I. (Ed) Berlin, Eidelberg: Springer, 1988.
4. Kruska, D. 'The effect of domestication on brain size and composition in the mink'. *J. Zool.* London, 1996; 239: 645-61.
5. Odent, M. 'Towards a super-brainy *Homo sapiens*?' In: Odent, M. *The Caesarean,* Free Association Books, London, 2004.
6. Odent, M., McMillan, L., Kimmel, T. 'Prenatal care and sea fish'. *European Journal of Obstetrics & Gynecology and Reproductive Biology* 1996; 68: 49-51.
7. Meeson, Lesley F. 'The effects on birth outcomes of discussions in early pregnancy, emphasizing the importance of eating fish'. PhD thesis, University of Wolverhampton, July 2007.
8. Tareke, E., Rydberg, P., Karlsson, P. et al. 'Analysis of acrylamide, a carcinogen formed in heated foodstuffs'. *J Agric Food Chem.* 2002;50 (17):4998-5006.
9. Pedersen, M., Von Stedingk, H., Botsivali, M. et al. 'Birth weight, head circumference, and prenatal exposure to acrylamide from maternal diet: the European prospective mother-child study (NewGeneris)'. *Environ Health Perspect* 2012 Dec;120(12):1739-45. doi: 10.1289/ehp.1205327. Epub 2012 Sep 23.
10. Kadawathagedara, M., Chan Hon Tong, A., Heude, B., et al. 'Dietary acrylamide intake during pregnancy and anthropometry at birth in the French EDEN mother-child cohort study'. *Environmental Research.* August 2016, 149:189-196.

Chapter 6: A turning point in maternal-foetal conflicts

1. Xiong, X., Demianczuk, N.N., Buekens, P., Saunders, L.D. 'Association of pre-eclampsia with high birthweight for age'. *Am J Obstet Gynecol.* 2000;183:148-155.
2. Chen, X.K., Wen, S.W., Smith, G., et al. 'Pregnancy-induced hypertension is associated with lower infant mortality in preterm singletons'. *BJOG* 2006; 113: 544-551.
3. Murphy, D.J., Sellers, S., MacKenzie, I.Z., et al. 'Case-control study of antenatal and intrapartum risk factors for cerebral palsy in very preterm singleton babies'. *Lancet* 1995;346:1449-54.
4. Siepmann, T., Boardman, H., Bilderbeck, A., et al. 'Long-term cerebral white and gray matter changes after preeclampsia'. *Neurology* 2017 Mar 28;88(13):1256-1264. doi: 10.1212/WNL.0000000000003765. Epub 2017 Feb 24
5. Lao, T.T., Chin, R.K., Swaminathan, R., Lam, Y.M. 'Maternal thyroid hormones and outcome of pre-eclamptic pregnancies'. *Br J Obstet Gynaecol* 1990 Jan;97(1):71-4
6. Narin, N., Kurtoğlu, S., Başbuğ, M., et al. 'Thyroid function tests in the newborn infants of pre-eclamptic women'. *J Pediatr Endocrinol Metab* 1999 Jan-Feb;12(1):69-73
7. Odent, M. 'Pre-eclampsia as a maternal-fetal conflict. The link with fetal brain development. *International Society for the Study of Fatty Acids and Lipids (ISSFAL) News.* 2000;7:7-10.
8. Wang, Y., Kay, H.H., Killam, A.P. 'Decreased levels of polyunsaturated fatty acids in pre-eclampsia'. *Am J Obstet Gynecol.* 1991;164:812-818.
9. Velzing-Aarts, F.V., van der Klis, F.R., van der Dijs, F.P., Muskiet, F.A. 'Umbilical vessels of pre-eclamptic women have low contents of both n-3 and n-6 long-chain polyunsaturated fatty acids'. *Am J Clin Nutr.* 1999;69:293-298.
10. Williams, M.A., Zingheim, R.W., King, I.B., Zebelman, A.M. 'Omega-3 fatty acids in maternal erythrocytes and risk of pre-eclampsia'. *Epidemiology.* 1995;6:232-237
11. Carlson, E., Salem, N. 'Essentiality of omega-3 fatty acids in growth and development in infants'. In: Simopoulos, A.P., Kifer, R.R., Martin, R.E., Barlow, S.M., (Eds) *Effects of Polyunsaturated Fatty Acids in Seafoods. World Rev Nutr Diet*; Basel, Karger; 1991; 66: 74-86.
12. Al, M.D., Van Houwelingen, A.C., Hornstra, G. 'Relation between birth order and the maternal and neonatal docosahexaenoic acid status'. *Eur J Clin Nutr.* 1997;51:548-553
13. Poston, L., Brilley, A.L., Seed, P.T., et al. 'Vitamin C and vitamin E in pregnant women at risk for pre-eclampsia (VIP trial): randomised placebo-controlled trial'. *Lancet* 2006;367:1145-54.
14. Eclampsia Trial Collaborative Group. 'Which anticonvulsant for women with eclampsia?' *Lancet.* 1995;345:1455-1463.
15. Bucher, H.C., Guyatt, C.H., Cook, R.J., et al. 'Effect of calcium supplementation on pregnancy-induced hypertension and pre-eclampsia'. *JAMA.* 1996;275:1113-1117.
16. Kiilhoma, P., Pakarinen, P., Gronroos, M. 'Copper and zinc in pre-eclampsia'. *Acta Obstet Gynecol Scand.* 1984;63:629-631.
17. Bodnar, L.M., Catov, J.M., Simham, H.N., et al. 'Maternal Vitamin D deficiency increases risk of pre-eclampsia'. *J. Clin. Endocrinol Metabol* 2007; 92(9):3517-22.
18. Olsen, S.F., Secher, N.J. 'A possible preventive effect of low-dose fish oil on early delivery and pre-eclampsia: indications from a 50-year-old controlled trial'. *Br J Nutr.* 1990;64:599-609
19. Odent, M. 'Gestational diabetes and Health Promotion'. *Lancet* 2009;374:684
20. Odent, M. 'The primary human disease. An evolutionary perspective'. *ReVision* 1995; 18 (2): 19-21
21. Odent, M. 'Hypothesis: Preeclampsia as a Maternal-Fetal Conflict'. *MedGenMed* September 5, 2001. © 2001 Medscape, Inc.

22. Odent, M. 'Obstetrical implications of the aquatic ape hypothesis.' In: *Was Man more Aquatic in the Past?*. Marc Verhaegen, Algis Kuliukas, Mario Vaneeechoutte (Ed.) Chapter 9. Bentham Science, 2009.
23. Chaline, J. 'Increased cranial capacity in hominid evolution and pre-eclampsia.' *J Reprod Immunol.* 2003 Aug;59(2):137-52.
24. Mendez, F.L., Poznik, G.D., et al. 'The divergence of Neandertal and Modern Human Y chromosomes.' *AJHG* April 2016; 98(4): 728-734.
25. Dannemann, M., Andres, A.M., Kelso, J. 'Introgression of Neandertal- and Denisovan-like Haplotypes Contributes to Adaptive Variation in Human Toll-like Receptors.' *Am J Hum Genet.* 2016 Jan 7;98(1):22-33. doi: 10.1016/j.ajhg.2015.11.015.
26. Sankararaman, S., Mallick, S., et al. 'The genomic landscape of Neanderthal ancestry in present-day humans.' *Nature* 2014 Mar 20;507(7492):354-7. doi: 10.1038/nature12961. Epub 2014 Jan 29.
27. Simonti, C.N., Vernot, B., et al. 'The phenotypic legacy of admixture between modern humans and Neandertals.' Science 12 Feb 2016:Vol. 351, Issue 6274, pp. 737-741 DOI: 10.1126/science.aad2149
28. McCoy, R.C., Wakefield, J., Akey, J.M. 'Impacts of Neanderthal-Introgressed Sequences on the Landscape of Human Gene Expression.' *Cell* 2017 Feb 23;168(5):916-927.e12. doi: 10.1016/j.cell.2017.01.038.

Chapter 7: A turning point in the attraction to water during labour

1. Michel Odent, *Water and Sexuality*, Arkana (Penguin) 1990.
2. Daniel Everett, *Don't sleep, there are snakes,* Profile Books, 2008.
3. Odent, M. 'La reflexotherapie lombaire. Efficacité dans le traitement de la colique néphrétique et en analgésie obstétricale', *La Nouvelle Presse Medicale* 1975; 4 (3):188.
4. Huntley, A.L., Coon, J.T., Ernst, E. 'Complementary and alternative therapies for labor pain: a systematic review', *Am J Obstet Gynecol,* 2004 Jul;191(1):36-44
5. Odent, M. 'Birth under water', *Lancet* 1983;2:1376-77.
6. Odent, M. 'Can water immersion stop labor?' *J Nurse Midwifery* 1997 Sep-Oct;42(5):414-6.
7. Odent, M. 'What I learned from the first hospital birthing pool', *Midwifery Today* 2000;54:16.
8. Odent, M. 'A landmark in the history of birthing pools', *Midwifery Today* 2000;54:17-8, 69.
9. Gilbert, R.E., Tookey, P.A. 'The perinatal mortality and morbidity among babies delivered in water', *BMJ* 1999; 319: 483-7.
10. Austin, T., Bridges, N., et al. 'Severe neonatal polycythaemia after third stage of labour under water', *Lancet* 1997 Nov 15:350:1445.

Chapter 8: A turning point in the microbial colonisation of newborn humans

1. Virella, G., Silveira Nunes, M.A., Tamagnini, G. 'Placental transfer of human IgG subclasses', *Clin Exp Immunol.* 1972 Mar;10(3):475-8.
2. Pitcher-Wilmott, R.W., Hindocha, P., Wood, C.B. 'The placental transfer of IgG subclasses in human pregnancy', *Clin Exp Immunol.* 1980 Aug;41(2):303-8.
3. Garty, B.Z., Ludomirsky, A., Danon, Y., Peter, J.B. and Douglas, S.D. 'Placental transfer of immunoglobulin G subclasses', *Clin Diagn Lab Immunol.* 1994 Nov;1(6):667-9.
4. Borghesi, J., Mario, L.C., Rodrigues, M.N., Favaron, P.O., Miglino, M.A. 'Immunoglobulin Transport during Gestation in Domestic Animals and Humans—A Review', *Open Journal of Animal Sciences*, 2014;4: 323-336.
5. Van Nimwegen, F.A., Penders, J., Stobberingh, E.E., et al. 'Mode and place of delivery,

gastrointestinal microbiota, and their influence on asthma and atopy'. *J Allergy Clin Immunol* 2011 Nov;128(5):948-55.e1-3.
6. Odent, M. 'Entering the world of microbes'. In: *The Caesarean,* Free Association Books, London, 2004: 58-61.
7. Aagaard, K., Ma, J., Antony, K.M., Ganu, R., Petrosino, J., Versalovic, J. 'The placenta harbors a unique microbiome', *Sci Transl Med.* 2014 May 21;6(237):237ra65. doi: 10.1126/scitranslmed.3008599.
8. Kort, R., Caspers, M., de Graaf, A. 'Shaping the oral microbiota through intimate kissing'. *Microbiome.* 2014.
9. Lif Holgerson, P., Harnevik, L., Hernell, O., et al. 'Mode of birth delivery affects oral microbiota in infants', *J Dent Res.* 2011 Oct;90(10):1183-8. Epub 2011 Aug 9.
10. Kuitunen, M., Kukkonen, K., Juntunen-Backman, K., et al. 'Probiotics prevent IgE-associated allergy until age 5 years in caesarean-delivered children but not in the total cohort', *J Allergy Clin Immunol.* 2009 Feb;123(2):335
11. Dominguez-Bello, M.G., De Jesus-Laboy, K.M., Shen, N., et al. 'Partial restoration of the microbiota of caesarean-born infants via vaginal microbial transfer', *Nat Med.* 2016 Mar;22(3):250-3. doi: 10.1038/nm.4039. Epub 2016 Feb 1.
12. Odent, M. 'The fetus ejection reflex', *Birth* 1987: 14:104-105.
13. Odent, M. 'The future of neonatal BCG', *Medical Hypotheses* 2016; 91: 34-36.
14. Odent, M. 'Future of BCG', *Lancet* 1999; 354: 2170.
15. Aronson, N.E., Santosham, M., Comstock, G.W., et al. 'Long-term Efficacy of BCG Vaccine in American Indians and Alaska Natives', *JAMA.* 2004;291(17):2086-2091. doi:10.1001/jama.291.17.2086.

Chapter 9: A turning point in the classifications of human births

1. Donnison, Jean, *Midwives and Medical Men,* Heinemann, London, 1977.
2. Glavind, J., Uldbjerg, N. 'Elective caesarean delivery at 38 and 39 weeks: neonatal and maternal risks', *Curr Opin Obstet Gynecol* 2015 Apr;27(2):121-7. doi: 10.1097/GCO.0000000000000158.
3. Condon, J.C., Jeyasuria, P., Faust, J.M., Mendelson, C.R. 'Surfactant protein secreted by the maturing mouse fetal lung acts as a hormone that signals the initiation of parturition', *Proc Natl Acad Sci USA* 2004 Apr 6;101(14):4978-83. Epub 2004 Mar 25.
4. Gao, L., Rabbitt, E.H., et al. 'Steroid receptor coactivators 1 and 2 mediate fetal-to-maternal signaling that initiates parturition', *J Clin Invest.* 2015 Jul 1;125(7):2808-24. doi: 10.1172/JCI78544. Epub 2015 Jun 22.
5. Hauth, J.C., Parker, C.R. Jr, MacDonald, P.C., Porter, J.C., Johnston, J.M. 'A role of fetal prolactin in lung maturation', *Obstet Gynecol.* 1978 Jan;51(1):81-8.
6. Varendi, H., Porter, R.H., Winberg, J. 'The effect of labor on olfactory exposure learning within the first postnatal hour', *Behav Neurosci.* 2002 Apr;116(2):206-11.
7. Odent, M. 'The early expression of the rooting reflex', *Proceedings of the 5th International Congress of Psychosomatic Obstetrics and Gynaecology,* Rome, 1977. London: Academic Press, 1977: 1117-19.
8. Odent, M. 'L'expression précoce du réflexe de fouissement'. In: *Les cahiers du nouveau-né* 1978; 1-2: 169-185.
9. Hermansson, H., Hoppu, U., Isolauri, E. 'Elective Caesarean Section Is Associated with Low Adiponectin Levels in Cord Blood', *Neonatology* 2014;105:172-174 (DOI:10.1159/000357178).
10. Bagci, S., Berner, A.L., et al. 'Melatonin concentration in umbilical cord blood depends on mode of delivery', *Early Human Development* 2012; 88(6):369-373.
11. Christensson, K., Siles, C., et al. 'Lower body temperature in infants delivered by caesarean section than in vaginally delivered infants', *Acta Paediatr* 1993;82(2):128-31.

12. Downes, K.L., Hinkle, S.N., Sjaarda, L.A., et al. 'Prior Prelabor or Intrapartum Caesarean Delivery and Risk of Placenta Previa' *Am J Obstet Gynecol*. 2015 www.ajog.org/article/S0002-9378(15)00005-8.
13. Prior, E., Santhakumaran, S., Gale, S., et al. 'Breastfeeding after caesarean delivery: a systematic review and meta-analysis of world literature', *Am J Clin Nutr*. 2012 May;95(5):1113-35. doi: 10.3945/ajcn.111.030254. Epub 2012 Mar 28.
14. Zanardo, V., Savolna, V., Cavallin, F., et al. 'Impaired lactation performance following elective caesarean delivery at term: role of maternal levels of cortisol and prolactin', *J Matern Fetal Neonatal Med*. 2012 Sep;25(9):1595-8. doi: 10.3109/14767058.2011.648238. Epub 2012 Feb 6.
15. Cabrera-Rubio, R., Collado, M.C., Laitinen, K., et al. 'The human milk microbiome changes over lactation and is shaped by maternal weight and mode of delivery', *Am J Clin Nutr*. 2012 Sep;96(3):544-51. doi: 10.3945/ajcn.112.037382. Epub 2012 Jul 25.
16. Azad, M.B., Konya, T., Maugham, H., et al. 'Gut microbiota of healthy Canadian infants: profiles by mode of delivery and infant diet at 4 months', *CMAJ* February 11, 2013 cmaj.
17. Dogra, S., Sakwinska, O., Soh, S., Ngom-Bru, C., Brück, W.M., Berger, B., Brüssow, H., Lee, Y.S., Yap, F., Chong, Y., Godfrey, K.M., Holbrook, J.D. 'Dynamics of infant gut microbiota are influenced by delivery mode and gestational duration and are associated with subsequent adiposity', *mBio* 2015 6(1):e02419-14. doi:10.1128/mBio.02419-14.
18. Simon-Areces, J., Dietrich, M.O., Hermes, G., et al. 'Ucp2 Induced by Natural Birth Regulates Neuronal Differentiation of the Hippocampus and Related Adult Behavior', *PLoS ONE*, 2012; 7 (8): e42911 DOI: 10.1371/journal.pone.0042911.
19. Tyzio, R., Cossart, R., Khalilov, I., Minlebaev, M., Hubner, C.A., Represa, A., Ben-Ari, Y., Khazipov, R. 'Maternal oxytocin triggers a transient inhibitory switch in GABA signaling in the fetal brain during delivery', *Science* 2006; 314: 1788-1792.
20. Levine, L.D., Sammel, M.D., Hirshberg, A., et al. 'Does stage of labor at time of caesarean affect risk of subsequent preterm birth?', *Am J Obstet Gynecol*. 2014 Sep 30. pii: S0002-9378(14)01020-5. doi: 10.1016/j.ajog.2014.09.035. [Epub ahead of print]
21. Hannah, M.E., Hannah, W.J., et al. 'Planned caesarean section versus planned vaginal birth for breech presentation at term: a randomised multicentre trial', *Lancet* 2000;356:1375-83.
22. Odent, M. 'Birth under water', *Lancet* 1983;2:1376-77.
23. Odent, M. 'The birthing pool test', *Midwifery Today* 2015;115:9-11.
24. Norsk, P., Epstein, M. 'Effects of Water Immersion on Arginine Vasopressin Release in Humans.' *J Appl Physiol* 1985 64 (1): 1–10.
25. Gutkowska, J., Antunes-Rodrigues, J., McCann, S. 'Atrial Natriuretic Peptide in Brain and Pituitary Gland.' *Physiol Rev* 1997 77 (2): 465–515.
26. Mukaddam-Daher, S., et al. 'Regulation of Cardiac Oxytocin System and Natriuretic Peptide during Rat Gestation and Postpartum', *J Endocrinol* 2002 175 (1): 211–16.
27. Gutkowska, J., et al. 'Oxytocin Releases Atrial Natriuretic Peptide by Combining with Oxytocin Receptors in the Heart', *Proc Natl Acad Sci USA* 1997 94 (21): 11704–09.

Chapter 10: A turning point towards the symbiotic revolution

1. Odent, M. *Genèse de l'homme écologique*, Epi. Paris, 1979.
2. Odent, M. *The Scientification of Love,* Free Association Books, London, 1999.
3. *Genesis* 1, 28.
4. *Exodus* 20, 17.
5. *Proverbs* 22, 15.
6. Kropotkin, Peter *Mutual Aid: a factor of evolution*, Will Jonson, ISBN: 978-1497333734

7. Sagan, D. and Margulis, L. *Origins of sex: three billion years of genetic recombination,* Yale University Press, 1986.
8. Béchamp, A. *Les microzymas dans leurs rapports avec les fermentations et la physiologie,* Association Française pour l'Avancement des Sciences, Nantes, 1875.
9. Béchamp, A. *Les Microzymas,* first edition 1883. Second edition: Centre International d'études Antoine Béchamp, Paris, 1990.

Chapter 11: A turning point in the history of orgasmic states
1. Everett, D. *Don't Sleep, there are Snakes,* Profile Books, 2008.
2. DeMeo, J. *Saharasia,* Orgone Biophysical Research Lab, Greensprings, Oregon, 1998.
3. Kitzinger, S. *Woman's Experience of Sex,* Penguin, 1983.
4. DeMeo, J. 'The geography of male and female genital mutilation'. In: *Sexual Mutilations,* George C. Denniston and Marilyn Milos, Eds: 1-15 Plenum Press, New York, 1997.
5. Odent, M. 'Colostrum and civilization'. In: *The Nature of Birth and Breastfeeding,* Bergin and Garvey, Westport CT, 1992.
6. Deutsch, H. *Psychoanalysis of the sexual functions of women,* Karnac Books, London, New York, 1991 (written 1923–24).
7. Kroll, U. 'A womb-centred life'. In: *Sex and God,* Linda Hurcombe (Ed.), Routledge and Kegan Paul, London, 1987, p102.
8. Thirleby, A. *Tantra; the Key to Sexual Power and Pleasure,* Jaico, Bombay, 1982.
9. Odent, M. *The Functions of the Orgasms: The Highways to Transcendence,* Pinter and Martin, London, 2009.

Chapter 12: A turning point in our relationship with time
1. Schlinzig, T., Johansson, S., Gunnar, A., et al. 'Epigenetic modulation at birth – altered DNA-methylation in white blood cells after Caesarean section', *Acta Paediatr.* 2009;98:1096–9.
2. Almgren, M., Schlinzig, T., Gomez-Cabrero, D., et al. 'Caesarean delivery and hematopoietic stem cell epigenetics in the newborn infant: implications for future health?' *Am J Obstet Gynecol.* 2014 Jun 18. pii: S0002-9378(14)00465-7. doi: 10.1016/j.ajog.2014.05.014.
3. Godfrey, K.M., Sheppard, A., Gluckman, P.D., Lillycrop, K.A., Burdge, G.C., McLean, C., Rodford, J., Slater-Jefferies, J., Garratt, E., Crozier, S.R., Emerald, B.S., Gale, C.R., Inskip, H.M., Cooper, C., and Hanson, M.A. 'Epigenetic gene promoter methylation at birth is associated with child's later adiposity', *Diabetes* 60: doi: 10.2337/db10-0979 (2011).
4. Wahl, S., Drong, A., Lehne, B., et al. 'Epigenome-wide association study of body mass index, and the adverse outcomes of adiposity', *Nature* December 2016. doi:10.1038/nature20784

Chapter 13: A turning point in the size of human groups
1. Dunbar, R.I.M. 'Neocortex size as a constraint on group size in primates', *Journal of Human Evolution.* 1992;22(6): 469–493.
2. Mendel, G. *La chasse structurale,* Payot, Paris, 1977.
3. Kelly, R.C. 'The evolution of lethal intergroup violence', *Proc Natl Acad Sci USA.* 2005 Oct 25;102(43):15294-8. Epub 2005 Aug 29.
4. Hultman, C.M., Sparen, P., Cnattingius, S. 'Perinatal risk factors for infantile autism', *Epidemiology.* 2002 Jul;13(4):417-23.
5. Yip, B.H., Leonard, H., et al. 'Caesarean section and risk of autism across gestational age: a multi-national cohort study of 5 million births', *Int J Epidemiol* 2016 Dec 25.

pii: dyw336. doi: 10.1093/ije/dyw336. [Epub ahead of print]
6. Glasson, E.J., Bower, C., et al. 'Perinatal factors and the development of autism: a population study', *Arch Gen Psychiatry*. 2004 Jun;61(6):618-27.
7. Getahun, D., Fasset, M.J., et al. 'Association of Perinatal Risk Factors with Autism Spectrum Disorder', *Am J Perinatol* 2017 Jan 31. doi: 10.1055/s-0036-1597624. [Epub ahead of print].
8. Gregory, S.G., Anthropolos, R., et al. 'Association of autism with induced or augmented childbirth in North Carolina Birth Record (1990–1998) and Education Research (1997–2007) databases', *JAMA Pediatr*. 2013 Oct;167(10):959-66. doi: 10.1001/jamapediatrics.2013.2904.
9. Seidman, D.S., Laor, A., et al. 'Long-term effects of vacuum and forceps deliveries', *Lancet* 1991 Jun 29;337(8757):1583-5.
10. Madsen, K.M., Hviid, A., et al. 'A population-based study of measles, mumps, and rubella vaccination and autism', *N Engl J Med*. 2002 Nov 7;347(19):1477-82.
11. Kenyon, S., Pike, K., et al. 'Childhood outcomes after prescription of antibiotics to pregnant women with preterm rupture of the membranes: 7-year follow-up of the ORACLE I trial', *Lancet* 2008 Oct 11;372(9646):1310-8. doi: 10.1016/S0140-6736(08)61202-7. Epub 2008 Sep 17.
12. Kenyon, S., Pike, K., et al. 'Childhood outcomes after prescription of antibiotics to pregnant women with spontaneous preterm labour: 7-year follow-up of the ORACLE II trial', *Lancet* 2008 Oct 11;372(9646):1319-27. doi: 10.1016/S0140-6736(08)61203-9. Epub 2008 Sep 17.
13. Krehbiel, D., Poindron, P., et al. 'Peridural anaesthesia disturbs maternal behaviour in primiparous and multiparous parturient ewes', *Physiology and Behavior*. 1987; 40: 463-72.
14. Lundbland, E.G., Hodgen, G.D. 'Induction of maternal-infant bonding in rhesus and cynomolgus monkeys after caesarian delivery', *Lab. Anim. Sci* 1980; 30: 913.

Chapter 14: A turning point in the history of natural selection

1. Peterson, C. *Psycho-social-cultural risk factors for breech presentation*. PhD dissertation, Department of Anthropology, University of South Florida, 2 July 2008.
2. Gelis, J. *L'Arbre et le Fruit*. Fayard, Paris, 1984.
3. Jordan, B. 'External cephalic version', *Women and Health* 1982; 7 (3-4): 83-101.
4. Everett, D. *Don't Sleep, there are Snakes*, Profile Books, 2008.
5. Nordtveit, T.I., Melve, K.K., et al. 'Maternal and paternal contribution to intergenerational recurrence of breech delivery: population based cohort study', *BMJ* 2008 Apr 19;336(7649):872-6. doi: 10.1136.
6. Liu, S., Liston, R.M., Joseph, K.S., et al. 'Maternal mortality and severe morbidity associated with low-risk planned caesarean delivery versus planned vaginal delivery at term', *CMAJ* 2007;176(4):455-60.
7. Hannah, M.E., Hannah, W.J., Hewson, S.A., et al. 'Planned caesarean section versus planned vaginal birth for breech presentation at term: a randomised multicentre trial', *Lancet* 2000; 256: 1375-83.
8. Englemann, G.J. *Labor Among Primitive Peoples*, J.H. Chambers & Co., St Louis, 1884.
9. Odent, M. 'Colostrum and civilization'. In: Odent, M. *The Nature of Birth and Breast-feeding*, Bergin & Garvey, 1992. 2nd ed 2003 *Birth and Breastfeeding*, Clairview.
10. Odent, M. 'Neonatal tetanus', *Lancet* 2008; 371:385-386.
11. Fildes, V.A. *Breasts, bottles and babies. A history of infant feeding*, Edinburgh University Press, 1986.

Chapter 15: A turning point at two and a half minutes before midnight

1. Bulletin of the Atomic scientists. *Board moves the clock ahead.* 26 January 2017.
2. Odent, M. *Genèse de l'homme écologique*, Epi. Paris, 1979.
3. Alvergne, A., Jenkins, C., Faurie, C., (Eds) *Evolutionary Thinking in Medicine*, Springer, 2016.
4. Morgan, E. *The Scars of Evolution*, Souvenir Press, London, 1990.
5. Laughon, S.K., Branch, D.W., Beaver, J., Zhang, J. 'Changes in labor patterns over 50 years'. *Am J Obstet Gynecol.* 2012 May;206(5):419.e1-9. Epub 2012 Mar 10.
6. Konrath, S.H., O'Brien E.H., Hsing. C. 'Changes in dispositional empathy in American college students over time: a

Concusion: Clarification: Who is *Homo*?

1. Schwartz, J.H., Tattersall, I. 'Defining the genus Homo'. *Science* 2015 Aug 28;349(6251):931-2. doi: 10.1126/science.aac6182. Epub 2015 Aug 27.
2. Reich, D., Green, R.E., et al. 'Genetic history of an archaic hominin group from Denisova Cave in Siberia', *Nature* 2010 Dec 23;468(7327):1053-60. doi: 10.1038/nature09710.
3. Reich, D., Patterson, N., et al. 'Denisova admixture and the first modern human dispersals into Southeast Asia and Oceania', *Am J Human Genet.* 2011 Oct 7;89(4):516-28. doi: 10.1016/j.ajhg.2011.09.005. Epub 2011 Sep 22.
4. Sankararaman, S., Mallick, S., et al. 'The Combined Landscape of Denisovan and Neanderthal Ancestry in Present-Day Humans', *Curr Biol.* 2016 May 9;26(9):1241-7. doi: 10.1016/j.cub.2016.03.037. Epub 2016 Mar 28.
5. Qin, P., Stoneking, M. 'Denisovan Ancestry in East Eurasian and Native American Populations', *Mol Biol Evol* 2015; 32 (10): 2665-2674.
6. Gokhman, D., Lavi, E., et al. 'Reconstructing the DNA methylation maps of the Neandertal and the Denisovan', *Science.* 2014 May 2;344(6183):523-7. doi: 10.1126/science.1250368. Epub 2014 Apr 17.
7. Huerta-Sanchez, E., Jin, X., et al. 'Altitude adaptation in Tibetans caused by introgression of Denisovan-like DNA', *Nature.* 2014 Aug 14;512(7513):194-7. doi: 10.1038/nature13408. Epub 2014 Jul 2.
8. Racimo, F., Gokhman, D., et al. 'Archaic adaptive introgression in TBX15/WARS2', *Mol Biol Evol* 2016 Dec 21. pii: msw283. doi: 10.1093/molbev/msw283. [Epub ahead of print].
9. Hoover, K.C., Gokcumen, O., et al. 'Global Survey of Variation in a Human Olfactory Receptor Gene Reveals Signatures of Non-Neutral Evolution', *Chem Senses.* 2015 Sep;40(7):481-8. doi: 10.1093/chemse/bjv030. Epub 2015 Jun 13.
10. Dannemann, M., Andres, A.M., Kelso, J. 'Introgression of Neandertal- and Denisovan-like Haplotypes Contributes to Adaptive Variation in Human Toll-like Receptors', *Am J Hum Genet.* 2016 Jan 7;98(1):22-33. doi: 10.1016/j.ajhg.2015.11.015.

Index